Oklahoma Notes

Basic-Sciences Review for Medical Licensure
Developed at
The University of Oklahoma at Oklahoma City, College of Medicine

Suitable Reviews for:
National Board of Medical Examiners (NBME), Part I
Medical Sciences Knowledge Profile (MSKP)
Foreign Medical Graduate Examination in the Medical Sciences (FMGEMS)

Oklahoma Notes

Biochemistry

Edited by
Thomas Briggs
Albert M. Chandler

With Contributions by
Thomas Briggs   Wai-Yee Chan
Albert M. Chandler   A. Chadwick Cox   Jay S. Hanas
Robert E. Hurst   Leon Unger   Chi-Sun Wang

Springer-Verlag
New York Berlin Heidelberg
London Paris Tokyo

Thomas Briggs, Ph.D.
Department of Biochemistry and Molecular Biology
Health Sciences Center
The University of Oklahoma at Oklahoma City
Oklahoma City, OK 73190
U.S.A.

Albert M. Chandler, Ph.D.
Department of Biochemistry and Molecular Biology
Health Sciences Center
The University of Oklahoma at Oklahoma City
Oklahoma City, OK 73190
U.S.A.

Library of Congress Cataloging in Publication Data
Biochemistry.
  (Oklahoma notes)

Printed and bound by Edwards Brothers, Ann Arbor, Michigan.
Printed in the United States of America.

9 8 7 6 5 4 3 2

ISBN 0-387-96341-3 Springer-Verlag New York Berlin Heidelberg
ISBN 3-540-96341-3 Springer-Verlag Berlin Heidelberg New York

Preface to the
Oklahoma Notes

In 1973, the University of Oklahoma College of Medicine instituted a requirement for passage of the Part I National Boards for promotion to the third year. To assist students in preparation for this examination, a two-week review of the basic sciences was added to the curriculum in 1975. Ten review texts were written by the faculty: four in anatomical sciences and one each in the other six basic sciences. Self-instructional quizzes were also developed by each discipline and administered during the review period.

The first year the course was instituted the Total Score performance on National Boards Part I increased 60 points, with the relative standing of the school changing from 56th to 9th in the nation. The performance of the class has remained near the national candidate mean (500) since then, with a mean over the 12 years of 502 and a range of 467 to 537. This improvement in our own students' performance has been documented (Hyde et al: Performance on NBME Part I examination in relation to policies regarding use of test. J. Med. Educ. 60:439–443, 1985).

A questionnaire was administered to one of the classes after they had completed the boards; 82% rated the review books as the most beneficial part of the course. These texts have been recently updated and rewritten and are now available for use by all students of medicine who are preparing for comprehensive examinations in the Basic Medical Sciences.

RICHARD M. HYDE, Ph.D.
Executive Editor

PREFACE

This book is intended to be a *review*. We assume the reader has already had a course in biochemistry. The book covers only the highlights and omits much detailed knowledge that is usually found in textbooks, to which the reader should turn for further reference. We hope those who are studying for medical national board examinations will find it particularly useful. To this end we have included, with most chapters, some multiple-choice questions whose answers can be found in the text.

Each chapter has been contributed by a colleague who is an expert in the field and/or an experienced teacher of the subject. The words are the authors' own, though we have done some editing in order to achieve a reasonably consistent format. Because of the multiple authorship, there is inevitably some unevenness in the depth of treatment of the various topics, but we accept responsibility for decisions on what to include and what to leave out. John W. Campbell, Ph.D., a graduate of this institution, executed the drawings. We did some figures ourselves, typed most of the text, and printed all of it on our PC's.

Future editions, should this one meet with success, will be revised periodically. We welcome comments and constructive suggestions.

Thomas Briggs

Albert M. Chandler

CONTENTS

Chapter 1
A. Chadwick Cox

Amino Acids and Proteins

1

Chapter 2
Wai-Yee Chan

Enzymes

18

Chapter 3
Thomas Briggs

Energetics and
Biological Oxidation

33

Chapter 4
Robert E. Hurst

Carbohydrates

46

Chapter 5
Albert M. Chandler

Amino Acid Metabolism

77

Chapter 6
Thomas Briggs

Porphyrins

100

Chapter 7
Chi-Sun Wang

Lipids

105

Chapter 8
Thomas Briggs

Steroids

127

Chapter 9
Leon Unger

Purines and Pyrimidines

138

Chapter 10
Jay S. Hanas

Nucleic Acids
and Protein Synthesis

149

Chapter 11
Wai-Yee Chan

Human Genetics:
Inborn Errors of Metabolism

175

Chapter 12
Thomas Briggs

Nutrition

195

1. AMINO ACIDS AND PROTEINS

A.C. Cox

I. AMINO ACIDS

The twenty amino acids that make up proteins are called the common amino acids. These are the only amino acids that are coded for in DNA and mRNA. They are:

ALAnine	**LEU**cine
ARGinine	**LYS**ine
ASparagi**Ne**	**MET**hionine
ASPartic acid	**PHE**nylalanine
CYSteine	**PRO**line
GLYcine	**SER**ine
GLutami**Ne**	**THR**eonine
GLUtamic acid	**TYR**osine
HIStidine	**TR**y**P**tophan
Iso**LEU**cine	**VAL**ine

The bolded, capital letters indicate the three-letter abbreviation for each.

A. Stereochemistry

Amino acids, except glycine, have the L absolute configuration and can rotate the plane of polarized light. The amino group is attached to the carbon adjacent to the carboxyl group, the α-carbon, so it is called an α-amino group. The common amino acids have an R group attached to the α-carbon and differ from each other in the nature of this R group.

B. Solubility

Amino acids can be grouped according to the interactions of their side chains with water. **Hydrophobic** (apolar) amino acids include Leu, Ile, Val, Met, Phe, Pro, Trp and Tyr. **Hydrophilic** (polar) amino acids include Asp, Asn, Cys, Glu, Gln, Ser, Thr, Arg, Lys and His. Ala and Gly are considered to be **Neutral**.

C. Non-coded Amino Acids and Derivatives of the Common Amino Acids

Several amino acids are modified after being incorporated into protein chains (post-translational modifications). Examples of these include Ser, Thr and Tyr which are sometimes phosphorylated on their hydroxyl groups. Phosphorylations of this kind are very important in enzyme regulation and in the mechanism of hormone action.

Other modifications include:

1. The hydroxylation of Lys and Pro residues in proteins found in connective tissues such as collagen and elastin.

2. The attachment of oligosaccharides to hydroxyl groups of Ser or Thr or to the amide group of Asn to form glycoproteins.

3. The formation of disulfide bridges between Cys residues either within the same polypeptide chain or between chains. This is a spontaneous, non-enzymatic process in contrast to the modifications mentioned previously.

4. The formation of γ-carboxyl glutamic acid residues found in proteins involved in blood clotting. The carboxylations require the participation of Vitamin K.

Unusual amino acids are found in peptides synthesized by enzymatic mechanisms not involving the ribosomal process. D-amino acids are found in various bacterial and fungal peptides. Other amino acids exist and have important roles; for example, γ-aminobutyric acid is a neurotransmitter.

D. Amino Acids as Ampholytes

The amino and carboxyl groups of amino acids are ionizable and behave as if they were on separate molecules. That is, the state of ionization for each group can be calculated from the Henderson- Hasselbalch Equation,

$$pH = pKa + \log \text{[base]/[acid]}$$

where each term represents the conjugate base and acid form of the same group. When the concentrations of the acidic and basic forms of the group are equal, the ratio becomes 1.0, its log = 0 and the pH = pKa. This is also the point of maximum buffering capacity.

In the case of amino acids having only one amino group and one carboxyl group per molecule, the net charge on the molecule is a function of the pH. At very low pH values the charge on the carboxyl group is neutralized and the amino group carries a positive charge. At high pH values the charge on the amino group is suppressed and the carboxyl group carries a negative charge. At a neutral pH, both groups are charged, amino group (+) and carboxyl group (-), with the net charge = 0. Under this condition the molecule will not migrate in solution in an electric field and is referred to as a "zwitterion". The pH at this point is referred to as the isoelectric point (pI).

Figure 1.1 illustrates each of these points. The following questions and answers refer to this figure.

Q. How many pKa values exist?

A. The same number as equivalents of acid or base required to titrate the groups-- 3.

Q. What are their pK values?

A. The same as the pH at each half-equivalence. In this example, approximately 2, 4 and 10.

Q. What is the value of the isoelectric point?

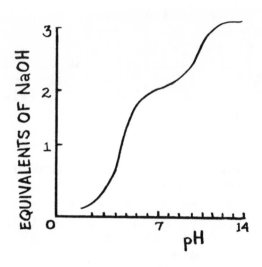

FIGURE 1.1

A. This is an acidic amino acid with two pKa values in the acidic region. The pI is halfway between the two acidic values, i.e., 3.0. Note that in this example, the titration curve in the basic region corresponds to the titration of the amino group and appears exactly as expected for the titration of any weak acid. Also note that the titration of the two acidic groups blends together in a continuous curve but since two equivalents of base are required it is concluded that two separate groups are present.

Q. At what pH or pH's is this compound a good buffer?

A. The answer is 3, because there are three ionizable groups present and each group is a good buffer at its pKa.

Several amino acids have other ionizable groups in addition to the groups attached to the α-carbon. These include:

Asp and Glu: a carboxyl group on the δ and γ carbons, respectively.
Arg: a guanidinium group that essentially remains positively charged at all physiological pH values.
His: an imidazole group that loses its positive charge near neutrality.
Lys: an ε amino group that loses its positive charge at about pH 9.
Tyr: a phenolic group that becomes negatively charged around pH 9.
Cys: a thiol group that becomes negatively charged at pH 9.

These side chain groups are free to ionize even when the amino acid is incorporated into proteins. The α amino and carboxyl groups are, of course, tied up in peptide bond formation. Sometimes the side chain groups are buried inside the folds of the protein and cannot ionize at their normal pKa values. Tyr is often in this category because it is not charged at neutral pH and prefers to be buried.

II. PROTEINS

Proteins are polymers of the twenty common amino acids (AAs) listed in the previous section. Each successive AA is joined to the next by an amide bond between the α-carboxyl and amino groups; this bond is called a **peptide bond**. By convention the AA with the free α-amino group is listed first in the sequence.

A. The Peptide Bond

The peptide bond has some important structural features that bear on the conformation of proteins.

1. **Planarity:** The peptide bond has double bond character because the electrons from the carbonyl oxygen are delocalized into the C-N bond. As with any double bond, the two bonded atoms, plus the four atoms attached to these two, are coplanar. Thus the α-carbon, the carbonyl carbon, the carbonyl oxygen, the amino nitrogen and the hydrogen it bears, and the next α-carbon in the chain, are all in the same plane. (In Figure 1.2 the atoms connected by the dotted lines and the atoms within the bounds of these lines are coplanar. Two separate peptide bonds are shown, hence two separate planes.)

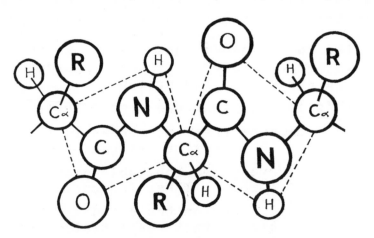

FIGURE 1.2

2. *Trans* **configuration:** There is little rotation around a double bond, thus there are two possible diastereomeric configurations, *trans* and *cis*. The peptide bond is much more stable in the trans form, the only form found in proteins. (In Figure 1.2 the peptide bonds are depicted in the trans configuration since the α-carbons are on opposite sides of the bond.)

3. **Restricted rotation:** Each α-carbon is part of the plane of the peptide bond occurring before and after it. However, each of these planes can rotate about the bonds attached to the α-carbon. (In Figure 1.2 the rotation of these two planes about an α-carbon is indicated by arrows and the angles labeled ϕ and psi.) The amount of rotation depends on what side chains (R groups) are present on the AAs on either side of the peptide bond. In general, the bulkier the groups,

the more restricted the rotation. Proline is a special case since the side chain is actually bonded back to the α amino group. This angle is fixed. The restriction of rotation, with this one exception, is due to steric factors.

    4. **Non-covalent bonds:** Many of the interactions between different parts of proteins are referred to as bonds but they are not of the co-valent type. They are:

        a. **Hydrogen bonds:** A hydrogen atom bonding to an electrophilic group such as oxygen or nitrogen will share itself with electrons on another electrophilic atom to which it is not formally bonded. The bond is electrostatic in nature but is directional along normal or-bital angles.

        b. **Salt bonds:** Also referred to as ionic bonds. These are elec-trostatic interactions between negatively and positively charged groups when they are very close together.

        c. **Hydrophobic bonds:** This term is commonly used but no formal bond exists. Apolar groups do associate with one another in aqueous solutions because energy is expended to organize water molecules about groups extending into the surrounding solution whereas less of the group is exposed to water in the associated form. The force in form-ing these associations is entropic in nature rather than enthalpic.

B. <u>Structural Levels of Proteins</u>

    1. **Primary Structure:** Primary structure refers to the AA sequence within a peptide chain including disulfide bonds and post-transla-tional modifications.

    2. **Secondary Structure:** Because the rotation of the peptide planes about the α-carbons are restricted, certain arrangements are much pre-ferred and are given special names. They are <u>right-handed α-helix</u>, <u>ß-pleated sheet</u>, <u>ß-turn</u> and <u>random structure</u>. The hydrogen bonds in the helix are parallel to the axis of the spiral, whereas in the pleated sheet the hydrogen bonds are between the adjacent peptide segments. If these segments have sequences running in the same direction (N to C), the structure is termed parallel; if the sequences are reversed, the structure is antiparallel. The ß-turn is the sharpest angle through which a peptide segment can be turned, and the first and fourth AAs of the turn are bonded to one another in antiparallel fashion.

    3. **Tertiary structure:** The overall three dimensional form of the protein constitutes the tertiary structure and is unique for each pro-tein. Not all proteins exhibit all four forms of the secondary struc-tures listed above. Some special proteins, like collagen, have other types of secondary structure.

    4. **Quaternary structure:** Many proteins are composed of several as-sociated subunits, where each subunit has its own tertiary structure and forms a higher order of structure termed the quaternary structure. Obviously only those proteins with subunits can have quaternary structure.

C. Protein Conformations

Under a given environmental condition a protein has a unique form although flexibility in the structure exists. However, if the environment is changed the protein may adopt another form. For example, many enzymes change form when they bind a substrate. These conformational changes are called **allosteric**.

Some proteins are secreted from cells in an inactive form. In most cases these proteins are proteolytically converted to their functional forms when they are needed. Blood coagulation proteins are a good example of this process and they are discussed below. The inactive form of these proteins are designated by either the prefix "pro", as in prothrombin or the suffix "ogen", as in fibrinogen or trypsinogen. If the active form is an enzyme then the inactive form is referred to as a **zymogen**. Trypsinogen is a zymogen but fibrinogen is not. In most cases the conversion of a zymogen to its active form requires the removal of a peptide referred to as the activation peptide, and leads to conformational changes.

In some cases, multimeric proteins may be composed of variable amounts of two or more closely related subunit forms. Such proteins are referred to as **isozymes**. Lactic dehydrogenase is a good example of an isozyme.

III. COLLAGEN AND FIBROUS PROTEINS

Several extracellular proteins are long, stringy molecules as opposed to the rather compact, folded form of most proteins. The former proteins are referred to as the fibrous proteins and collagen, the most plentiful of all proteins, is one example.

Collagen has a trimeric structure, where each polypeptide has a helical form and all of these helices are wound around each other. The helix of collagen is different from an α helix. There are four types of collagen that are distinguished by having polypeptides of different amino acid sequences but the same helical structure.

All types of collagen have three post-translational modifications. Two of these modifications are the hydroxylation of Lys and Pro residues. The hydroxylation of Pro increases the stability of collagen. These hydroxylations require **Vitamin C**. This is the reason that vitamin C deficiency (scurvy) affects the skin and connective tissues. Hydroxylysines are often glycosylated in the collagens.

Lys is also involved in the third modification. The ε amino group on the side chain is removed oxidatively leaving the ε carbon as an aldehyde. Two of these aldehyde groups can then condense into an aldol bond forming a covalent bridge between polypeptide segments that contain the two modified Lys residues. These bridges can link two parts of the same collagen molecule or two different molecules. Such intra and intermolecular couplings are a common feature of fibrous proteins.

Collagen contains a repeated amino acid sequence and this repeated sequence has a Gly residue at every third position. This is required because the helical form of the polypeptides brings every third amino acid so close to the other two polypeptides that there is only enough room for a hydrogen atom between them. This restricts this third amino acid to a Gly.

The three-stranded helical molecule is called tropocollagen before it polymerizes into a staggered three dimensional array cross linked together by aldol bonds and many hydrogen bonds. Collagen is synthesized in an inactive form called procollagen which is proteolytically processed outside the cell into tropocollagen which then polymerizes.

Another example of a fibrous protein is elastin. Its structure is not so highly extended nor as tightly repeated. In its polymerized form, the different helical forms are covalently cross-linked by a bond composed of four Lys residues called desmosine. Because the structure is more relaxed when these cross-links are formed, the overall polymerized elastin array can be stretched, but will return to its original form after the force is removed.

Collagen and elastin are the major components of structures such as tendons and ligaments. Other important fibrous proteins involved in linking different cells together are laminin and fibronectin. These two proteins are more like beaded chains in that several domains are linked together by flexible segments of polypeptide. Each domain binds to sites on different molecules found in the extracellular matrix located on most cells. For example, fibronectin contains domains that bind to collagen as well as others that bind to hyaluronic acid and heparin.

IV. HEMOGLOBIN AND OXYGEN TRANSPORT

In mammals, the delivery of oxygen to tissues requires a carrier system because the solubility of O_2 in water is too low to supply the need. The system consists of hemoglobin (Hb) carried in the erythrocytes. Hb binds O_2 and increases the oxygen-carrying capacity of blood about ten-fold. In addition, the carrier system provides a method for regulating the transport of O_2. In many tissues myoglobin serves to take the oxygen from hemoglobin and to act as a storage site and intracellular transport mechanism.

A comparison between myoglobin and hemoglobin is shown below:

Myoglobin	Hemoglobin
monomer	tetramer, 2 α and 2 β
binds one O_2	Binds $4O_2$, one/subunit
binding kinetics: hyperbolic	binding kinetics: sigmoidal
My + $O_2 \longrightarrow$ My.O_2	Hb + $4O_2 \longrightarrow$ Hb.$(O_2)_4$

The binding of oxygen to myoglobin as a function of the concentration of oxygen (partial pressure for gases) follows Michaelis-Menten kinetics. This is shown in Figure 1.3:

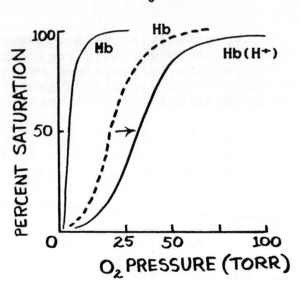

FIGURE 1.3

Hemoglobin has a very different reaction stoichiometry and the binding curve is also different, resembling that of an allosteric enzyme (Figure 1.3). One can write an equation for the equilibrium constant as follows:

$$K = [Hb(O_2)_4]/[Hb][O_2]^4$$

The binding curve plotted for this equation would still be sigmoidal, like the one indicated for hemoglobin in Figure 1.3, but would be more marked in its rise. The curve for hemoglobin fits an equation where the power term is not four as indicated in the equation above but 2.8 because the binding of oxygen is not perfectly cooperative as implied by the 4th power term.

The first oxygen that binds to hemoglobin causes a shift in the tertiary structure of the subunits of hemoglobin to a conformation that binds oxygen more readily. The conformation of hemoglobin in the deoxygenated state is referred to as the **tense state** and after the first oxygen molecule is bound the conformation is converted to the **relaxed state**. This new conformation state binds oxygen better, but not so well as to yield the 4th power term of perfect cooperativity. Note that myoglobin cannot exhibit cooperativity because it only has the one binding site for oxygen per each molecule of myoglobin. When oxygen is bound, myoglobin changes configuration, but cannot bind other oxygen molecules because other sites are not available.

The first oxygen molecule is bound to a **ferrous** atom associated with a planar heme group of one of the subunits. Four out of the six coordinate bonds that can be formed by a ferrous atom are bonded to the heme group, leaving two unfilled bonding orbitals, one on either side of the plane of the heme ring. One of these bonds is occupied by a histidyl residue of hemoglobin. The other is occupied by O_2 when this subunit of hemoglobin is oxygenated. When this site is not occupied, the ferrous atom projects slightly above the plane of the heme ring because of the asymmetry of forces. When oxygen binds to the

ferrous atom these forces bring the ferrous atom back into the plane
of the heme ring. Although this shift is very small, it is magnified
in other parts of the molecule by a lever effect causing a more significant structural change in the subunit which in turn causes changes
in the other subunits associated with it. Thus a small shift in the
position of the oxygen molecule is transferred into relatively large
conformational changes in the total hemoglobin molecule. These changes allow the three remaining subunits to bind oxygen more readily.

Oxygen binding by hemoglobin is an exquisitely regulated system.
Various other factors influence oxygen binding, for example, H^+
concentration.

$$Hb(H^+)_2 + 4O_2 \rightleftharpoons Hb(O_2)_4 + 2H^+$$

As indicated in this equation, changes in hydrogen ion concentration lead to changes in oxygen binding. The system acts as an
acid/base buffer as well as a mechanism for improving the delivery of
oxygen to peripheral tissues. In the capillary beds of peripheral
tissues CO_2 is released as a byproduct of metabolism. After hydration
the CO_2 forms carbonic acid leading to a decrease in pH, i.e., an increase in H^+. According to the equation above this will lead to a release of O_2. As long as H^+ remains bound to the discharged hemoglobin
molecule the latter remains in the tense state and the binding of the
first O_2 molecule is inhibited. Thus, once deoxygenated in the presence of H^+ the molecule remains deoxygenated.

The process is reversed when the blood returns to the lung. In the
lung, CO_2 is expired and, therefore, the H^+ concentration is reduced
(pH increased). This favors the relaxed state of the hemoglobin
molecule which is now capable of binding oxygen again and the binding
curve again shifts back to the left (Figure 1.3). The release of hydrogen ions from hemoglobin upon oxygenation is called the **Bohr effect**, after its discoverer.

Carbon dioxide also binds to the deoxygenated molecule and has an
allosteric effect like that of pH. In the lung more O_2 is bound because the CO_2 concentration is decreased. In the capillaries where CO_2
concentration is high, O_2 is released. The combined effect of an increase in H^+ and CO_2 concentrations causes even a greater release of
O_2 than either one alone.

One other compound has an allosteric effect on hemoglobin. **2,3-
Diphosphoglycerate (DPG)** also binds to and stabilizes deoxygenated
hemoglobin, one molecule between two ß subunits. It does not bind to
oxygenated hemoglobin at the normal oxygen tension of the lung. At
the low oxygen tension of tissues, however, DPG binds, favoring deoxygenated Hb, so that O_2 dissociates more readily. Thus the efficiency
of oxygen transport is enhanced by the reversible binding of DPG. DPG
increases when lung O_2 concentration decreases as in hypoxia or at
high altitudes.

Mutations are known to affect about every aspect of hemoglobin
function from the protection of the ferrous state of the iron atom to
the form of polymerization of hemoglobin in the red blood cell. The

best known mutation occurs in the **Sickle Cell** state. The substitution of <u>Val</u> for <u>Glu</u> at the sixth position of the <u>ß subunit</u> causes hemoglobin in the deoxygenated state to form a very stable polymer that distorts the cell to an extent that it can block a capillary. Although this has dire consequences for the homozygote, heterozygous individuals have a mixed population of hemoglobin molecules that do not polymerize to block capillaries. Heterozygous individuals are also protected from malaria. In regions where malaria has killed a large percentage of the population over the centuries, heterozygotes have increased in percentage of that population.

More than one form of hemoglobin occurs in man. Not only is there the major form of hemoglobin referred to as adult or "A", but other hemoglobins are present in smaller amounts. Fetal hemoglobin, the predominant form in the fetus, decreases rapidly in amount after birth. Fetal hemoglobin binds oxygen more tightly than adult hemoglobin so that the fetus can extract oxygen from its mother's blood supply in an efficient manner.

IV. PLASMA PROTEINS

Plasma contains many non-enzyme proteins. Some of the blood coagulation proteins and the immunoglobulins fit in this category. Other proteins bind very specifically to small molecules and are referred to as transport proteins. The lipoproteins are one example of this kind. A few transport proteins are listed below.

Albumin: Albumin is a very non-specific transporter that binds to a large number of organic ions including many drugs.

Transferrin: carries two ferric atoms and delivers them to those cells having receptors for transferrin.

Retinol—Binding Protein: carries retinol and is complexed with thyretin which transports thyroxin.

Testosterone/Estradiol— Binding Protein: One of a group of rather specific steroid-binding proteins present in plasma and interstitial fluid.

Haptoglobin: binds heme released from dead cells and returns the iron back into metabolic circulation. Also protects the kidneys from harmful levels of heme.

Carrier proteins:

1. Increase the water solubility of hydrophobic molecules, i.e., lipids.

2. Decrease the loss of small molecules in the kidney. Many transported molecules would be filtered out in the kidneys unless a specific mechanism existed for taking them back up into the bloodstream. Transport molecules are, therefore, usually much larger than 50 kD so that they will not be filtered out themselves. Retinol-binding protein is an exception but it complexes tightly to the much larger <u>thyretin</u>.

3. Target the bound molecules to a particular tissue. These target tissues usually have very specific receptors.

4. Aid in detoxification, in the sense of preventing harmful amounts of potentially toxic molecules from accumulating in free form. For example, albumin binds lysophosphatidyl choline and free fatty acids.

Albumin is a particularly complicated carrier because of its rather non-specific binding properties. The most important molecules it transports are fatty acids, especially during fatty acid mobilization from adipose tissue. During such mobilization, the fatty acids compete for the same sites occupied by other compounds, including drugs, and in certain instances can cause a sharp rise in the free drug concentration as a result of competitive displacement.

Albumin is by far the most abundant plasma protein, especially on a molecular basis. The **osmotic pressure** associated with plasma is mainly due to albumin and if the albumin concentration drops it usually leads to edema because there is insufficient osmotic force to balance the arterial pressure.

Some proteins found in plasma at trace levels are not purposefully secreted but arise from lysed, dead cells. Many are enzymes that are easy to assay and are very specific for certain tissues and diagnostic for that tissue's damage.

V. HEMOSTASIS AND BLOOD COAGULATION

Hemostasis, the stopping of blood flow at a wound site, is brought about by the combined effort of platelets, the vessel wall, and plasma coagulation factors. The chronology of hemostasis is as follows. A wound exposes the collagen layer just below the endothelial cell layer that lines the vessel wall. Platelets recognize and bind to the collagen, become activated, recruit other platelets to form a hemostatic plug, secrete many substances that promote coagulation and support coagulation once it starts.

Platelets recruit other platelets by secreting **ADP**, a platelet agonist and by producing **thromboxane A$_2$**, another very potent platelet agonist. The synthesis of thromboxane from **arachidonic acid**, an essential fatty acid, requires cyclooxygenase, an enzyme blocked by aspirin. **Aspirin**, therefore, inhibits hemostasis by blocking platelet function.

Thromboxane A$_2$ and **serotonin**, another secreted compound, cause the smooth muscle cells of the vessel wall to contract, thus helping in hemostasis.

Platelets and the damaged tissues provide initiators of the coagulation process. Platelets also provide a phospholipid surface necessary for several of the reactions involved in coagulation.

The plasma coagulation proteins are numbered by roman numerals. A subscripted "a" indicates that a factor has been converted to its "active" form. **Prothrombin**, Factor II (F II), **thrombin**, its active form, and **fibrinogen**, F I, are almost always still referred to by their trivial names.

The exact reaction sequence *in vivo* is not known, but two partial, separate reaction series occur *in vitro* depending on how the reaction is initiated. The **intrinsic series** is initiated by activation of F XII by contact with glass. F XIIa then activates F XI which in turn activates F IX. F IXa along with its helper protein, F VIII, activates F X, which in turn with its helper protein, F Va, activates prothrombin. Thrombin then cleaves fibrinogen which spontaneously polymerizes to form the clot. The activations are all proteolytic cleavages and the activating factor is a <u>protease</u>. The helper proteins aid in holding the substrate factors in place during the activation and each of these helper proteins is activated proteolytically by thrombin. The two reactions involving helper proteins require a phospholipid surface for proper reaction velocities. **Calcium ions** are required for all these reactions.

The **extrinsic reaction** series is initiated by the addition of tissue factor, a helper protein containing its own phospholipid component. It helps F VII to activate F X; the rest of the reaction pathway is the same as for the intrinsic system after the activation of F X. Calcium ions are required for these reactions also.

The intrinsic and extrinsic reactions form the basis for diagnostic tests, e.g., prothrombin time and partial thromboplastin time. The former tests for for Factors I, II, V, VIII, IX, X, XI and XII; the latter for deficiencies in Factors I, II, VII AND X. Platelet function is tested separately by the bleeding time.

In the physiological reaction scheme, platelets supply the phospholipid surface for the F IX and F X-catalyzed steps. While bound with their helper proteins to the platelets, they are protected from inhibition by the plasma protease inhibitors, **antithrombin III** and **α-macroglobulin**. This scheme keeps the coagulation process restricted to the wound site since the activating reactions are bound to platelets attached to the wound site.

Thrombin is the first protease that is freed from the platelets in a viable, active form. It causes the clot. Any thrombin that escapes from the area of the clot is rapidly bound to the intact endothelial cell surface and either inhibited or it activates an anticoagulation scheme.

Thrombin binds to **thrombomodulin**, its receptor on endothelial cells which also binds to **protein C**. Thrombomodulin acts like a helper protein in the thrombin activation of protein C , which in turn with its helper protein, **protein S**, inhibits F Va and F VIIIa. This is the reaction mechanism that ensures that if an activated platelet escapes the hemostatic plug, the coagulation process will be terminated. Furthermore, thromboxane causes endothelial cells to produce **prostaglandin I₂** which is a potent antagonist of platelet activation.

Fibrin, the product of the cleavage of peptides A and B from fibrinogen by thrombin action, polymerizes on its own, but as with other fibrous proteins, fibrin polymers are covalently cross-linked.

F XIIIa, a <u>transglutaminase</u>, joins the side chains of a glutamine to a lysine on a different fibrin molecule. Platelets bind fibrin and help in organizing the fibrin strands and even pull these into a tighter form to maintain a rigorous clot. Thrombin activates F XIII.

The proteases of this scheme, including protein C, have modified glutamic acid residues. They have an extra carboxyl group on the γ carbon and are, therefore, referred to as γ-carboxyl glutamic acid residues (Gla). These post-translationally modified residues bind the proteases to the phospholipid surface via Ca^{++} bridges. The post-translational modification requires a **vitamin K-dependent step**, therefore, vitamin K deficiency blocks coagulation. **Warfarin** and other coumarin compounds mimic and block vitamin K action and also inhibit coagulation. These drugs are used routinely to prevent recurrent thrombosis, an abnormal coagulation that occurs on atherosclerotic vessel walls. Several other approaches have been attempted to prevent thrombosis. Drugs that block platelet action, such as aspirin, have only a limited benefit in managing thrombosis.

The fibrin clots are removed proteolytically by **plasmin**. Plasmin is activated on the clots by **plasminogen activator**, a protein released from tissues as a response to the clotting process. This leaves plasmin on the clot until it finishes the hydrolysis. An anti-plasmin inhibitor protein is present in plasma to scavenge any plasmin that escapes from the clot prematurely.

Another approach used to treat thrombosis is to activate the fibrinolysis reactions. Plasminogen activator; **urokinase**, a plasminogen activator isolated from urine; and **streptokinase**, a bacterial activator are all used to treat thrombosis on an acute basis.

Another method used to control thrombosis is through the use of **heparin**, an anticoagulant. Heparin acts as a catalyst for the inactivation of thrombin by antithrombin III, the plasma protein inhibitor. Heparin must be injected for this treatment.

Blood, after removal from a subject is anticoagulated most often by using calcium ion chelators, like citrate or EDTA. These cannot be used *in vivo* because the Ca^{++} concentration in blood is very critical for many processes.

VI. REVIEW QUESTIONS ON AMINO ACIDS AND PROTEINS

DIRECTIONS: Each of the questions or incomplete statements below is followed by five suggested answers or completions. Select the one that is BEST in each case and fill in the corresponding space on the answer sheet.

The figure below refers to questions 1 and 2:

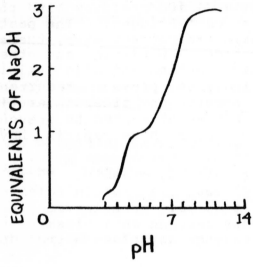

The curve below refers to question 3:

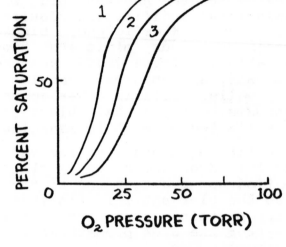

1. Which one of the following amino acids can fit the curve shown?

A. Ser
B. Asp
C. Pro
D. His
E. Gly

2. How many charges are on the molecule at pH 4?

A. one negative charge
B. half the molecules have one negative charge and half have no net charge
C. half the molecules have one positive charge and half have no net charge
D. half the molecules have two positive charges and half have one positive charge
E. two positive charges

3. Given that the curve in the figure represents the O_2 binding curve for hemoglobin, an increase in the CO_2 concentration but without a change in the pH would cause the curve to:

A. shift to curve 1
B. remain the same
C. shift to curve 3
D. remain in the same position as curve 2 but increase in magnitude
E. none of the above

4. The effect of a normal dose of aspirin would be to increase:

A. bleeding time
B. partial thromboplastin time
C. prothrombin time
D. all of the above
E. none of the above

5. Many proteins contain two or more segments (domains) displaying homology in sequence. These proteins have evolved by:

A. post-translational modification
B. epigenetic alteration
C. gene duplication and fusion
D. allosteric cooperativity
E. transpeptide formation

6. Which one of the following groups of amino acids are all hydrophobic?

A. Thr, Asp, Glu
B. Leu, Lys, Val
C. Ile, Leu, Lys
D. Val, Leu, Ala
E. Arg, Ile, Val

7. Hemophilia, caused by a deficiency in Factor VIII, exhibits an increase in:

A. bleeding time
B. partial thromboplastin time
C. prothrombin time
D. B and C
E. A, B and C

8. The binding of 2,3-diphosphoglycerate to hemoglobin occurs primarily between:

A. the alpha subunits in the oxygenated state
B. each alpha-beta subunit pair in the oxygenated state
C. each alpha-beta subunit pair in the deoxygenated state
D. the beta subunits in the deoxygenated state
E. none of the above

9. The differences in the binding curves between hemoglobin and myoglobin are due primarily to:

A. the amount of oxygen that can be carried by the different heme groups
B. the markedly different tertiary structures
C. subunit interactions
D. the differences in pH of the surrounding solution
E. lysyl substitution for the histidyl group

DIRECTIONS: For each of the questions or incomplete statements below, ONE or MORE of the answers or completions is correct. On the answer sheet fill in space

    A   if only 1, 2, and 3 are correct
    B   if only 1 and 3 are correct
    C   if only 2 and 4 are correct
    D   if only 4 is correct
    E   if all are correct

FILL IN ONLY ONE SPACE ON YOUR ANSWER SHEET FOR EACH QUESTION

Directions Summarized				
(A)	(B)	(C)	(D)	(E)
1,2,3	1,3	2,4	4	All are
only	only	only	only	correct

10. The endothelial cell may be considered the principal anticoagulant cell because it:

1. produces prostaglandin I_2
2. secretes plasminogen activator
3. supports protein C activation
4. consumes protein S

11. Thrombin, the first free protease produced by the coagulation cascade, activates:

1. Factor V
2. Factor VIII
3. Factor XII
4. protein C

12. Which one of the following pairs represents an inhibitor and what it inhibits?

1. prostaglandin I_2 -- platelet activation
2. thromboxane A_2 -- endothelial cell activation
3. protein C - protein S -- Factor VIIIa
4. α_2-macroglobulin -- Factor Va

13. Zymogens are converted into active enzymes by:

1. enzyme adenylate formation
2. phosphorylation
3. ATP activation
4. limited proteolysis

14. The common forces responsible for tertiary structure are:

1. hydrogen bonds
2. ionic bonds
3. hydrophobic interactions
4. steric hindrance

15. The Henderson-Hasselbalch equation can be used to calculate the:

1. pH of a solution if the pKa and the ratio of conjugate acid to conjugate base are known
2. pKa if the hydrogen ion, conjugate base and the total weak acid concentrations are known
3. ratio of base to acid if the Ka and hydrogen ion concentration are known
4. pKa when the concentration of the conjugate base is exactly equal to that of the conjugate acid

16. An example of a covalent crosslink found in a fibrous protein is:

1. desmosine
2. lysyl aldol bond
3. glutaminyl-lysine transpeptide bond
4. salt bridge

17. Ascorbic acid, needed for the synthesis of functional collagen, is required for reactions catalyzed by:

1. lysine hydroxylase
2. lysine oxidase
3. proline hydroxylase
4. glucose transferase

FILL IN ONLY ONE SPACE ON YOUR ANSWER SHEET FOR EACH QUESTION

		Directions Summarized		
(A) 1,2,3 only	(B) 1,3 only	(C) 2,4 only	(D) 4 only	(E) All are correct

18. Albumin is:

1. the major carrier of sterols
2. a transporter of fatty acids
3. the most basic of the major plasma proteins
4. the major protein contributor to the plasma osmotic force

19. The serum retinol binding protein

1. Transports the lipophilic retinol molecule through the aqueous medium of serum
2. is necessary for retinol uptake by target cells
3. protects membranes from retinol damage
4. binds to transferrin

20. The three individual polypeptide helices of collagen are tightly wound about each other at each turn in a super helix which requires:

1. lysines to be hydroxylated
2. lysines to be oxidized
3. prolines to be hydroxylated
4. periodicity in glycine positions.

21. In which of the following cases would the amount of oxygen carried by the blood be reduced?

1. anemia
2. hyperventilation
3. uncompensated acidosis
4. heterozygous sickle cell trait

VII. ANSWERS TO QUESTIONS ON AMINO ACIDS AND PROTEINS

1. D	8. D	15. A
2. D	9. C	16. A
3. C	10. A	17. B
4. A	11. E	18. C
5. C	12. B	19. A
6. D	13. D	20. D
7. C	14. E	21. B

2. ENZYMES

Wai-Yee Chan

I. NATURE OF ENZYMES

A. Introduction

1. All known enzymes are **proteins**.

2. Enzymes are biological **catalysts**, produced by living tissues, that increase the **rates** of reactions.

B. Definition of Terms

1. **Substrate**: substance acted upon by an enzyme.

2. **Activity**: amount of substrate converted to product by the enzyme per unit time (e.g. micromoles/minute).

3. **Specific activity**: activity per quantity of protein (e.g. micromoles/minute/mg protein).

4. **Catalytic constant**: proportionality constant between the reaction velocity and the concentration of enzyme catalyzing the reaction. Unit: activity/mole enzyme.

5. **Turnover number**: catalytic constant/number of active sites/mole enzyme.

6. **International Unit (IU)**: quantity of enzyme needed to transform 1.0 micromole of substrate to product per minute at 30°C and optimal pH.

C. Nomenclature

1. Trivial names, e.g. pepsin, trypsin, etc.

2. Addition of suffix **—ase** to name of substrate, e.g. arginase, which catalyzes conversion of arginine to ornithine and urea.

3. Systematic names; 6 major classes:

   a. **Oxidoreductase**: oxidation-reduction reactions, e.g., alcohol:NAD oxidoreductase for the enzyme catalyzing the reaction $RCH_2OH + NAD^+ \rightleftharpoons RCHO + NADH + H^+$.

   b. **Transferase**: transfer of functional groups including amino, acyl, phosphate, one carbon, glycosyl groups etc. Example: ATP: Creatine phosphotransferase for the enzyme catalyzing the reaction ATP + Creatine $\rightleftharpoons$ Phosphocreatine + ADP.

   c. **Hydrolase**: Cleavage of bond between carbon and some other atoms by the addition of water. Example: Peptidase for the enzyme catalyzing $R_1CONHR_2 + H_2O \rightleftharpoons R_1COOH + R_2NH_2$.

   d. **Lyase**: Add or remove the elements of water, ammonia or CO_2 to or from a double bond. Example: Phenylalanine ammonia lyase for the enzyme catalyzing the reaction phenylalanine $\rightleftharpoons$ cinnamic acid + ammonia.

   e. **Isomerase**: Racemization of optical or geometric isomers.

      i. epimerase or racemase for optical isomers (asymmetric carbon), as in D-lactic acid $\rightleftharpoons$ L-lactic acid (racemase).

      ii. mutase for geometric isomers or intramolecular group transfer, as in 2-phosphoglycerate $\rightleftharpoons$ 3-phosphoglycerate.

   f. **Ligase**: Formation of C-O, C-S, C-N, and C-C with the hydrolysis of ATP. Example: Pyruvate carboxylase for the enzyme catalyzing the reaction pyruvate + ATP + CO_2 $\rightleftharpoons$ oxaloacetate + ADP + P_i.

D. Basic Enzyme Structures

1. Enzymes may be composed of a single **polypeptide chain**, or several identical or different **subunits**.

2. Some compounds (organic or inorganic) other than amino acid side chains may be required for activity and are not modified at end of reaction:

   a. **prosthetic group**: when tightly bound to enzyme, as in heme of cytochrome.

   b. **cofactor**: when less tightly bound, or removable by dialysis, e.g. metal ions.

3. **Coenzymes**: organic molecules fulfilling the role of substrate, being modified at end of reaction, but readily regenerated by another linked reaction. Examples: biotin, NAD, ATP, TPP, FAD, pyridoxal phosphate, coenzyme A, etc.

4. **Holoenzyme**: Catalytically active enzyme complex consisting of a protein apoenzyme and a non-protein cofactor.

5. **Zymogen** (Proenzyme): Activatable precursor of enzyme. Active mature enzyme is generated by specific cleavage of a peptide bond. Examples: chymotrypsinogen $\longrightarrow$ chymotrypsin; pepsinogen $\longrightarrow$ pepsin.

E. Characteristics of Enzymatic Reactions

1. Enormous **catalytic power**.

2. Can be **saturated** by an excess of substrate.

3. Reaction **velocity** is directly **proportional** to enzyme **concentration**, provided there is enough substrate. The enzymatic reaction continues until substrate is exhausted.

4. **Optimum temperature**: the temperature at which the enzyme activity is at its maximum.

   a. Enzymes from different tissues of the same organism do not necessarily have the same temperature profile.

b. Rate of most enzymatic reactions about doubles for each 10ºC rise in temperature.

c. Temperature at which the reaction rate is maximum varies according to conditions such as salt content, pH, etc.

5. **Optimum pH:** pH at which the enzyme activity is at its maximum.

   a. optimum pH depends on the acid-base behavior of the enzyme and substrate.

   b. optimum pH of enzyme varies widely; the majority of enzymes have optima between 4 and 8.

   c. K_m and V_{max} vary independently at different pH's.

   d. Enzymes are denatured at extreme pH or temperature.

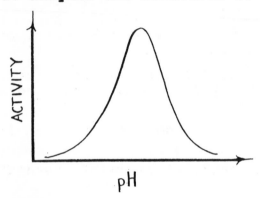 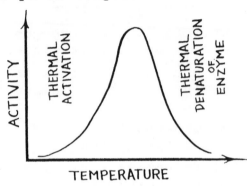

6. High **substrate specificity** can be absolute (one particular compound), or more broadly applicable to a class of compounds sharing a type of linkage, steric structure (cis-trans), or optical activity (D,L).

7. Activity is regulated by **feedback inhibition**, altered **availability of substrate**, or altered **kinetic parameters**.

II. ENZYME KINETICS

A. Basic Principles

1. **Order** of Reaction: if $A \rightleftharpoons P$, then $v = k[A]^R$, where

   v = velocity of reaction
   k = rate constant
   R = order of reaction
   [A] = concentration of reactant

2. **Energetics** of Catalysis (see figure, next page)

   a. A---B **transition state** (activated complex): the rate of reaction depends on the number of activated molecules in the transition state.

   b. E_A, or ΔG^* or ΔF^*: **free energy of activation**, the amount of energy which must be put into the system to reach the activated transition state.

c. ΔG or ΔF: **free energy change** of reaction is the difference in free energy of the reactants and products.

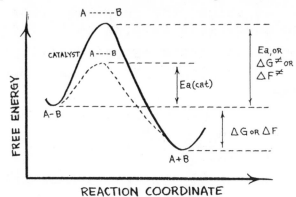

d. Catalysts form transition complexes with **lower energy of activation, E_A(cat)**.

3. **Equilibrium constant**

$$K_{eq} = \frac{[P_1] \times [P_2] \times \ldots}{[A_1] \times [A_2] \times \ldots}$$

The value of K_{eq} is **NOT** changed by the catalyst.

4. Enzyme **lowers the E_A for** both the forward and back reactions, therefore the velocity for both reactions is faster and equilibrium is achieved sooner.

B. **Enzyme Kinetics**

1. **Michaelis-Menten** Equation: given the reaction

where

$$E + S \underset{k_2}{\overset{k_1}{\rightleftharpoons}} ES \xrightarrow{k_3} E + P$$

E = enzyme
S = substrate
ES = enzyme-substrate complex
P = product
ET = total enzyme
k_1, k_2, k_3 = rate constants

then $v = \dfrac{V_{max}(S)}{K_m + (S)}$

a. V_{max}: maximal velocity achieved when enzyme is saturated with substrate.

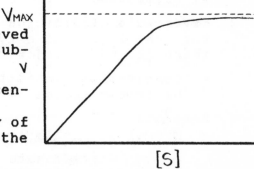

i. V_{max} is proportional to concentration of enzyme.
ii. measures catalytic efficiency of enzyme: the bigger the V_{max} the more efficient the enzyme.

b. K_m: Michaelis constant. Equal to the substrate concentration at which the reaction rate is half of its maximal value; units of moles/liter.
   i. $K_m = (k_2 + k_3)/k_1$

ii. Measures **catalytic power** of enzyme, the smaller the Km the greater the catalytic power and the more specific the enzyme

iii. High K_m indicates weak binding between enzyme and substrate; when dissociation of ES complex to E and P is the rate limiting step (i.e. k_1, $k_2 \gg k_3$) K_m becomes dissociation constant of ES.

c. **Lineweaver-Burk** plot

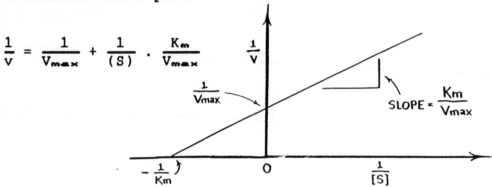

$$\frac{1}{v} = \frac{1}{V_{max}} + \frac{1}{(S)} \cdot \frac{K_m}{V_{max}}$$

C. Enzyme Inhibition

Two major types: irreversible and reversible. Drugs are designed to inhibit specific enzymes in specific metabolic pathways.

1. **Irreversible** inhibition: involves destruction or covalent modification of one or more functional groups of the enzyme. Examples:

    a. Diisopropylfluorophosphate and other fluorophosphates bind irreversibly with the -OH of the serine residue of acetylcholine esterase.

    b. Para-chloromercuribenzoate reacts with the -SH of cysteine.

    c. Alkylating agents modify cysteine and other side chains.

    d. Cyanide and sulfide bind to the iron atom of cytochrome oxidase.

    e. Fluorouracil irreversibly inhibits thymidine synthetase.

2. **Reversible** Inhibition: characterized by a rapid equilibrium of the inhibitor and enzyme, and obeys Michaelis-Menten kinetics. There are three major types:

    a. **Competitive** inhibition: resembling the substrate, the inhibitor competes with it for binding to the active site of the enzyme.

       Examples: | Inhibitor | Enzyme Inhibited |
       |-----------|------------------|
       | malonate | succinate dehydrogenase |
       | sulfanilamide | dihydropteroate synthetase |
       | methotrexate | dihydrofolate reductase |
       | Allopurinol | xanthine oxidase |

b. **Noncompetitive** inhibition: inhibitor does not resemble substrate, binds to enzyme at locus other than the substrate
binding site. Examples: heavy metal ions: silver, mercury,
lead, etc. Metalloenzymes are inhibited by metal-chelating
agents that bind metal cofactors, e.g. EDTA.

c. **Uncompetitive** inhibition: inhibitor binds to the enzyme-substrate complex and prevents further reaction.

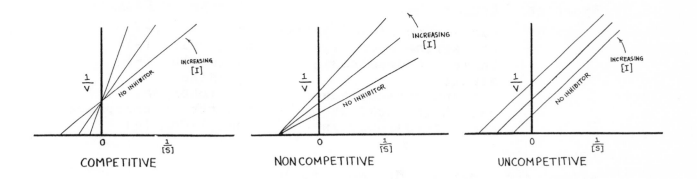

D. Active Site

1. The **active site** of an enzyme is the region that binds the substrate and contributes the amino acid residues that directly
participate in the making and breaking of bonds.

2. It is a three-dimensional entity.

3. Two types of amino acid residues: the **contact** (or catalytic)
amino acids and the **auxiliary** amino acids.

III. EFFECT OF KINETIC PARAMETERS ON ENZYME ACTIVITY

A. Allosteric Enzymes

1. General Properties

a. Enzyme activity is modulated through the noncovalent binding
of a specific metabolite (allosteric effector, modulator or
modifier) to the protein at the regulatory site **other than**
the catalytic site.

b. All known allosteric enzymes have 2 or more polypeptide **subunits,** often four.

c. Binding of the modulator to the allosteric site affects the binding of substrate to the catalytic site by changing the **quaternary structure** of the allosteric enzyme. The effect can be either **positive**, i.e. increases the binding of substrate; or **negative**, i.e. decreases the binding of substrate.

d. Regulation frequently occurs at the first, or **committed step** of a metabolic pathway, or at a **branch point**, with the final product of the pathway as a negative effector. This is end product or feedback inhibition.

Examples:	Enzyme	Allosteric Effector
	homoserine dehydrogenase	threonine (-)
	homoserine succinylase	methionine (-)
	threonine deaminase	isoleucine (-)
	aspartate transcarbamoylase	cytidine triphosphate (-)
	phosphofructokinase	fructose-6-phosphate (+)
	pyruvate carboxylase	acetyl-CoA (+)

2. Kinetics

a. Allosteric enzymes do **not** follow Michaelis-Menten kinetics.

b. A **sigmoidal** rather than hyperbolic curve is obtained when reaction rate is plotted against substrate concentration.

c. Kinetic behavior is analogous to oxygen binding of hemoglobin. Oxygen binding to myoglobin follows Michaelis-Menten kinetics instead.

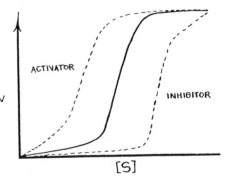

B. <u>Covalently Modified Regulatory Enzymes</u> (Phosphorylated Regulatory Enzymes)

Covalent binding of modifier (phosphate) to enzyme modifies enzyme activity. Example: phosphorylation of active glycogen synthetase turns the enzyme into the inactive form, while phosphorylation of inactive glycogen phosphorylase turns the enzyme into the active form (See Chapter 4, pages 56 - 58).

IV. GENERAL ASPECTS

A. Isozymes

1. Catalyze the **same reactions** and have the same molecular weight but **differ** in **subunit composition** and physical chemical properties.

2. **May differ** in K_m, V_{max}, optimal temperature and pH, substrate specificity.

3. Often contain **multiple polypeptide subunits** of 2 or more types.

4. Example: L-Lactate Dehydrogenase (LDH)

   a. **tetramer** of 2 types of subunit, M (muscle) and H (heart)

   b. five isozymes: H_4, H_3M, H_2M_2, HM_3, and M_4. Various mixtures of isozymes are found in different tissues.

   c. H_4 predominantly in cardiac tissue, and M_4 in skeletal muscle and liver.

B. <u>Medical Aspects of Enzymology</u>

  1. **As Diagnostic Tools**

    a. Enzymes or isozymes normally found intracellularly in various organs can be used as indicators of organ damage when they are found in blood. Examples:

Condition	Enzymes with Elevated Levels in Blood
Myocardial Infarction	glutamic-oxaloacetic transaminase (SGOT) lactic dehydrogenase H_4 and H_3M isozymes creatine phosphokinase
Bone disease	alkaline phosphatase
Obstructive liver disease	sorbitol dehydrogenase lactic dehydrogenase M_4 and M_3H isozymes
Prostatic cancer	acid phosphatase
Acute pancreatitis	amylase
Muscular dystrophy	aldolase glutamic pyruvate transaminase (SGPT)

    b. These enzymes although always present in blood at low levels, are elevated far above normal in pathological conditions.

  2. **As Laboratory Reagents**

    a. **Simple enzyme assays**: Enzymes may be used for accurate determination of small quantities of blood constituents.

    b. **Coupled enzyme assays**: combinations of enzymes are often used to measure concentrations of specific substrates, coenzymes, or products. For example, an enzyme reaction may be coupled to the conversion of NADH to NAD^+, the removal or production of which can be followed easily by measuring the absorbance of the solution spectrophotometrically.

    c. If the assay is for substrate (or product), then enzyme, coenzymes, etc. must be in excess; if the assay is for enzyme, then substrate, etc. must be in excess.

V. REVIEW QUESTIONS ON ENZYMES

DIRECTIONS: Each of the questions or incomplete statements below is followed by five suggested answers or completions. Select the one that is BEST in each case and mark the corresponding space on the answer sheet.

The diagram below refers to Questions 1-3:

The conversion of fumarate to malate is catalyzed by fumarase. The following results were obtained when the reaction was studied in the presence or absence of the competitive inhibitor malonate:

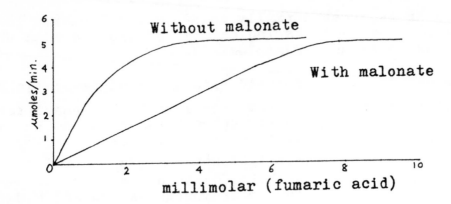

1. In the absence of inhibitor the V_{max} is

A. 1 μmole/min
B. 2 μmole/min
C. 3 μmole/min
D. 4 μmole/min
E. 5 μmole/min

2. In the absence of inhibitor the K_m for fumarate is

A. 0.5 millimolar
B. 1 millimolar
C. 1.5 millimolar
D. 2 millimolar
E. 4 millimolar

3. The effect of malonate

A. is to increase ΔG of the reaction.
B. is to decrease the V_{max} to 2.5 μmole/min
C. is to increase the K_m to 6 millimolar
D. can be overcome by increasing the concentration of fumarate.
E. results in cooperative binding of fumarate.

4. The effect of a catalyst on a reaction is to:

A. increase the energy of activation.
B. decrease the energy of activation.
C. increase the ΔG of the reaction.
D. decrease the ΔG of the reaction.
E. increase the equilibrium constant.

5. Lactic dehydrogenase is a tetrameric enzyme of 2 types of subunits, H and M, which associate according to the tissue of origin. The number of different tetrameric forms possible is

A. 1
B. 2
C. 3
D. 4
E. 5

6. Lactic dehydrogenase isozymes differ in

A. molecular weight
B. K_m for pyruvate
C. number of subunits
D. A and B are correct
E. A and C are correct

7. Data that would be useful in developing an assay for a serum enzyme of clinical interest include all of the following EXCEPT:

A. the molecular weight of the pure enzyme
B. K_m values for substrates so that they can be added to the asssay in appropriate amounts.
C. the occurrence of activation or inhibition by the substrate.
D. the stability of the enzyme under various conditions.
E. The optimal pH of the enzyme.

Questions 8-9:

The following data were obtained for an enzyme:

[Substrate] in M	Reaction velocity (μmole product/min)
1×10^{-4}	10
2×10^{-4}	20
3×10^{-4}	30
6×10^{-4}	55
9×10^{-4}	60
12×10^{-4}	60
15×10^{-4}	60

8. The V_{max} is

A. 10
B. 20
C. 30
D. 55
E. 60

9. The K_m is about

A. 1×10^{-4}M
B. 2×10^{-4}M
C. 3×10^{-4}M
D. 4.5×10^{-4}M
E. 6×10^{-4}M

DIRECTIONS for questions 10-15: The graph below contains several lines, each marked with a letter which represents a possible answer to a question. For each question, select the ONE BEST answer and fill in the appropriate space on the answer sheet.

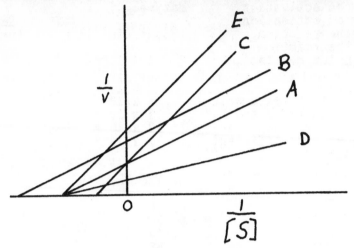

Questions 10-15: Line A represents the Lineweaver-Burk plot for the reaction of a normal substrate in the absence of any inhibitor for a given concentration of enzyme.

10. Which line would be expected in the presence of an uncompetitive inhibitor?

11. Which line would be expected in the presence of a noncompetitive inhibitor?

12. Which line would be expected in the presence of a competitive inhibitor?

13. Which line would be expected if the concentration of enzyme is doubled?

14. Which line has the same V_{max} as Line A but a smaller K_m?

15. Which line has the same K_m as Line A but a smaller V_{max}?

DIRECTIONS: For each of the questions or incomplete statements below, ONE or MORE of the answers or completions is correct. On the answer sheet fill in space

A if only <u>1, 2, and 3</u> are correct
B if only <u>1 and 3</u> are correct
C if only <u>2 and 4</u> are correct
D if only <u>4</u> is correct
E if <u>all</u> are correct

FILL IN ONLY ONE SPACE ON YOUR ANSWER SHEET FOR EACH QUESTION

Directions Summarized				
(A) 1,2,3 only	(B) 1,3 only	(C) 2,4 only	(D) 4 only	(E) All are correct

16. In the graph on the right, the X-axis can be:

1. Concentration of enzyme.
2. pH.
3. Concentration of substrate.
4. Temperature

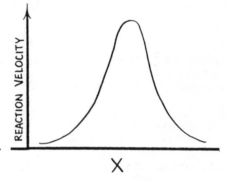

17. Urease is an enzyme following Michaelis-Menten kinetics. It catalyzes the reaction

$$urea + 2H_2O \rightleftharpoons 2NH_3 + H_2CO_3$$

in plants, and is inhibited by glycine. In the diagram at the right, which of the following materials could have been added at time T?

1. More glycine.
2. More urease.
3. More ammonia.
4. More urea.

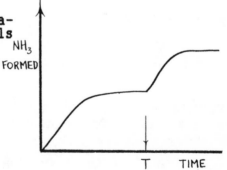

18. The effect of pH on an enzyme catalyzed reaction reflects

1. ionization of the enzyme-substrate complex.
2. Ionization of the enzyme.
3. Ionization of the substrate.
4. Denaturation of the enzyme.

19. Enzymes synthesized as zymogens include

1. carboxypeptidase.
2. chymotrypsin.
3. pepsin.
4. trypsin

20. Conversion of zymogen to active enzyme involves

1. phosphorylation.
2. methylation.
3. dimer formation.
4. cleavage of one or more specific peptide bonds.

21. The rate at which products are formed in a biosynthetic pathway can depend on the concentration of

1. substrate.
2. enzyme.
3. coenzyme.
4. inhibitor.

FILL IN ONLY ONE SPACE ON YOUR ANSWER SHEET FOR EACH QUESTION

Directions Summarized				
(A) 1,2,3 only	(B) 1,3 only	(C) 2,4 only	(D) 4 only	(E) All are correct

22. In catalyzing a reaction, an enzyme

1. increases the reaction velocity in both directions.
2. shifts the equilibrium toward product formation.
3. lowers the energy of activation.
4. minimizes the change in free energy.

23. The oxygenation of hemoglobin results in

1. change in the absorption of visible light.
2. change in the distance between heme groups.
3. increase in the acidity of some group in hemoglobin.
4. decreased binding of 2,3-diphosphoglycerate.

24. The active site of an enzyme

1. contains both contact and auxiliary amino acids.
2. provides an environment favorable for the binding of the substrate and the enzyme.
3. is maintained in proper conformation by the total 3-dimensional structure of the enzyme molecule.
4. may contain, in addition to amino acid side chains, nonprotein constituents essential for catalytic action of the enzyme.

DIRECTIONS: Each question below presents two entities which are to be compared quantitatively. On the answer sheet fill in space

  A if A is larger than B
  B if B is larger than A
  C if A and B are equal, or nearly equal
  D if the relative values of A and B cannot be determined from the information given

25. A K_m if affinity for substrate is high
B K_m if affinity for substrate is low

26. A enzymatic activity of an apoenzyme
B enzymatic activity of a holoenzyme

27. A V_{max} if catalytic efficiency of an enzyme is high
B V_{max} if catalytic efficiency of an enzyme is low

28. A enzymatic activity of a zymogen
B enzymatic activity of a mature enzyme

DIRECTIONS: The group of items below consists of five lettered headings, followed by several numbered questions. For each numbered question select the BEST lettered heading and fill in the corresponding space on the answer sheet. Each heading may be used once, more than once, or not at all.

Questions 29-32: Blood or serum levels of enzymes primarily found intracellularly in various organs are used as indicators of organ damage. For each of the diseases named below select the lettered enzyme with which it is most closely associated.

A. Glutamic-oxaloacetic transaminase
B. Glutamic-pyruvic transaminase
C. Alkaline phosphatase
D. Acid phosphatase
E. Sorbitol dehydrogenase

29. Bone disease

30. Muscular dystrophy

31. Myocardial infarction

32. Obstructive liver disease

Questions 33-38:

A. Reaction velocity
B. Concentration of substrate
C. Concentration of enzyme
D. Temperature
E. pH

(Questions 33-38
are the numbered axes
in the graphs below)

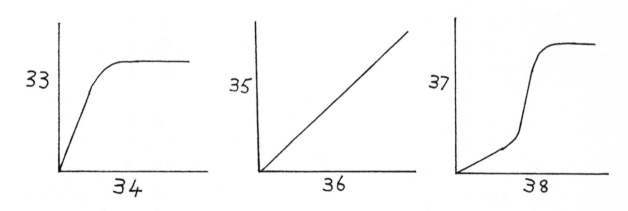

Questions 39-42:

A. Diisopropylfluorophosphate
B. Parachloromercuribenzoate
C. Sulfanilamide
D. Fluorouracil
E. Cyanide

39. Inhibitor of thymidine synthetase.

40. Inhibitor of acetylcholine esterase.

41. Inhibitor of enzyme with catalytically essential sulfhydryl group.

42. Inhibitor of cytochrome oxidase.

Questions 43-46:

A. Biotin
B. NAD
C. FAD
D. Pyridoxal phosphate
E. GTP

43. Pyruvate carboxylase

44. Lactate dehydrogenase

45. Glutamic-oxaloacetic transaminase

46. Amino acid oxidase

VI. ANSWERS TO QUESTIONS ON ENZYMES

1. E		24. E	
2. B		25. B	
3. D		26. B	
4. B		27. A	
5. E		28. B	
6. B		29. C	
7. A		30. B	
8. E		31. A	
9. C		32. E	
10. B		33. A	
11. E		34. B	
12. C		35. A	
13. D		36. C	
14. C		37. A	
15. E		38. B	
16. C		39. D	
17. D		40. A	
18. E		41. B	
19. E		42. E	
20. D		43. A	
21. E		44. B	
22. B		45. D	
23. E		46. C	

3. ENERGETICS AND BIOLOGICAL OXIDATION

Thomas Briggs

I. THERMODYNAMICS

For a reaction $A + B \rightleftharpoons C + D$ at equilibrium, the **equilibrium constant** is

$$K_{eq} = \frac{[C]\ [D]}{[A]\ [B]}$$

ΔG is the observed change in **free energy**, the useful energy produced by a reaction, the energy available to do useful work. Sometimes (incorrectly) called ΔF. If the reaction starts at standard conditions of pH 7 and 1 $\underline{M}$ concentrations of all reactants and products, and goes to equilibrium, then the change in free energy is the **standard** free energy change (denoted by superscript zero and prime) and is related to K_{eq} (R is the gas constant; T, the absolute temp.):

Eq. (1) $\qquad \Delta G^{o\prime} = -RT \ln K_{eq}$

A reaction with an equilibrium constant >1 will have a **negative** $\Delta G^{o\prime}$, and will tend to go to the right, or **down an energy hill**. But this tendency may be reversed by **concentration** differences or input of **energy** from somewhere else. A reaction can be driven against an unfavorable equilibrium if it is coupled, through a common intermediate, with an energy-releasing reaction.

A "spontaneous" reaction may not actually proceed all by itself, due to the presence of an activation energy barrier. One function of enzymes is to lower this barrier and speed up **rate** of reaction. The final position of equilibrium is related to $\Delta G^{o\prime}$ and **not** affected by enzymes. At equilibrium, $\Delta G = 0$.

Entropy is the degree of **disorder**, or randomness. Entropy is higher in a denatured than in a native protein. Entropy of living systems is kept low (highly ordered) at the expense of an increase in entropy of the surroundings. ΔS is the entropy change associated with a particular transformation. A spontaneous reaction **can** proceed with a decrease in entropy, if the **total** entropy change, including surroundings, is positive.

II. OXIDATION AND REDUCTION

Oxidation can be defined in three ways:
   1. add oxygen;
   2. remove hydrogen;
   3. **remove electrons**: this is the most general definition.
Reduction is the converse of the above.

Biological oxidation and reduction are always linked. For every oxidation, there **must** be a reduction; for every electron donor, there **must** be an electron acceptor.

Reducing power, E, also called **redox potential**, is the capacity to donate electrons. It is measured, in volts, by comparing with a standard hydrogen electrode.

$\Delta E_o'$ is the difference in **standard** redox potential (standard conditions) between electron donor and acceptor. In an oxidation-reduction reaction, the component with the more negative E_o' is the stronger electron donor or reducing agent, and will tend to reduce (be oxidized by) the second component. In Eq. (2), $\mathcal{F}$ is the faraday, a constant; n is the number of electrons transferred:

Eq (2) $\Delta E_o' = \dfrac{RT}{n\mathcal{F}} \ln K_{eq}$

A large K_{eq} (i.e., reaction tends to proceed far to right) is associated with a large **positive** $\Delta E_o'$.

Combining Eq (1) and Eq (2) we relate $\Delta E_o'$ and $\Delta G_o'$:

Eq (3) $\Delta G_o' = -n\mathcal{F}\Delta E_o'$

Points to remember:
1. a reaction that goes "spontaneously" has a **negative** ΔG.
2. a reaction that goes "spontaneously" has a **positive** ΔE.
3. a substance with the more **negative** redox potential will reduce another with a less negative or more positive redox potential:
     electrons tend to flow from **E negative to E positive**.
4. the **molecular oxygen** system, toward which electrons flow, has the most positive redox potential.
5. the directional tendency that a reaction might normally have **may** be reversed by:
     a. sufficient **concentration** difference (mass action)
     b. input of sufficient **energy**

III. ELECTRON TRANSFER VIA THE RESPIRATORY CHAIN

In catabolism, the usual electron (and H) acceptor is NAD^+. Flavoproteins are also used: FAD, FMN-- protein-bound. Reduced coenzymes ($FMNH_2$, $FADH_2$, NADH + H^+) **must be reoxidized**. In aerobic metabolism, this is by the electron transport system.

Electron Transport System (ETS): a chain of enzymes, located on the **inner mitochondrial membrane**, specialized to carry out electron transfer from reduced coenzymes to oxygen. Though separate, they are arranged in such a way that each receives an electron from the one just before it in the sequence, and in turn reduces the one next in line. At each transfer of an electron, a drop in free energy (ΔG is negative) occurs. At three points, sufficient free energy is released to drive the phosphorylation of ADP to ATP, a "high-energy" compound.

The Mitochondrion: about 1 x 3 μm (Figure 3.1). A liver cell has about 1000. The inner membrane is **very selective** in its **permeability**, and contains the enzymes of electron transport. Cristae are extensions of the inner membrane. The matrix has most enzymes of the citric acid cycle.

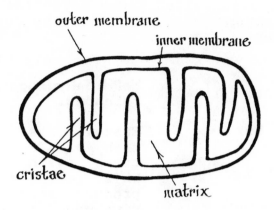

Figure 3.1. The Mitochondrion

Several enzymes of the ETS are **cytochromes**-- heme proteins. In the cytochrome system, electrons are transferred because the valence of the iron can change from $Fe^{++} \rightleftharpoons Fe^{+++}$:

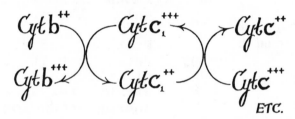

ETC.

Electron is transferred from cyt b to cyt c_1; cyt b is oxidized and cyt c_1 is reduced.	Now electron is transferred from cyt c_1 (reduced) to cyt c. Cyt c_1 is re-oxidized; cyt c is reduced.

<u>Coupled Oxidative Phosphorylation</u>: the process of ATP generation is tightly coupled to the process of electron transfer. If one stops, the other stops, like gears that mesh. This is the normal state in respiring mitochondria.

<u>Uncoupling</u>: some chemicals, e.g., 2,4-dinitrophenol (DNP) can uncouple phosphorylation, allowing e^- transfer to oxygen to proceed. Energy is wasted as heat; no ATP is formed. Like depressing the clutch on a car, engine runs but no useful work is done.

<u>P/O Ratio</u>: the number of ATP's formed per atom of O consumed in metabolism of substrate. NADH has P/O of 3; $FADH_2$, 2 (see scheme). An uncoupled system has P/O of zero for all substrates.

<u>Regulation</u>: depends on **ATP/ADP ratio** and on availability of P_i. If ATP predominates, the cell doesn't need energy, and e^- transfer slows. If the cell needs energy, ADP predominates and is available for coupled phosphorylation, which now speeds up, allowing increased e^- transfer, re-oxidation of reduced coenzymes, and speeding up of Krebs (TCA) cycle. Availability of O_2, the terminal e^- acceptor, also limits, since lack of an acceptor would stop everything (Figure 3.2).

Mitochondrial Electron Transport System

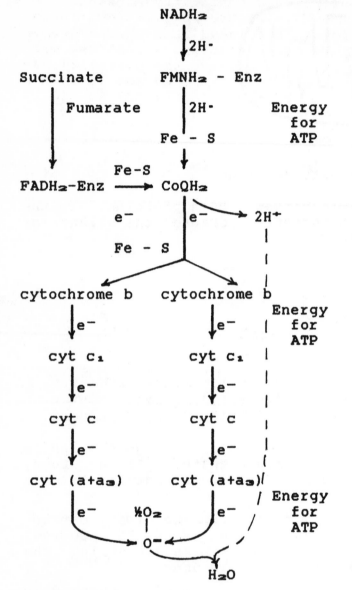

Comments

NADH dehydrogenase. Inhibited by rotenone.

Energy conserved here for generation of ATP while a pair of electrons (with 2H's) is transferred. This is the **first of 3 sites** that produce energy for coupled oxidative phosphorylation.

Coenzyme Q is a quinone (ubiquinone). Electrons from succinate enter ETS here. Since first energy-producing site is bypassed, $FADH_2$ from succinate can produce only 2 ATP's.

Protons and electrons now separate, but it still takes **a pair** of electrons to generate enough energy for 1 ATP here, the **2nd energy-producing site.** Antimycin A inhibits here.

All cytochromes contain a protein plus heme (i.e., iron protoporphyrin) or a heme derivative.

Also called **cytochrome oxidase.** Contains Cu as well as Fe. This is the 3rd **energy-producing site.** Inhibited by cyanide (CN^-).

Energy for ATP: e^- transfer expels H^+ from mitochondrion. Reentry drives phosphorylation by **chemi-osmotic coupling** (page 37).

Figure 3.2. Mitochondrial Electron Transport

IV. CHEMI-OSMOTIC THEORY OF OXIDATIVE PHOSPHORYLATION

The components of the electron transport system are highly ordered
in the inner mitochondrial membrane, such that electron transfer results in a **directional extrusion of H⁺** (Figure 3.3) from inside
(matrix) to outside the mitochondrion. This leads to a **pH difference**
(more acid outside) and an **electrochemical gradient** across the membrane, a condition of potential energy. To release the potential energy, protons are allowed back in through the **F₁ - F₀ ATPase** in such a
way (details unknown) as to drive the phosphorylation of ADP to ATP.

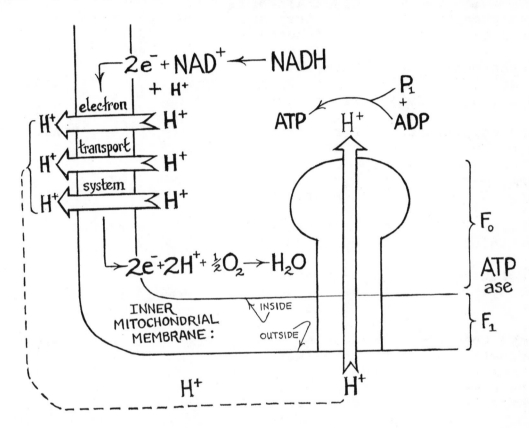

Figure 3.3. Oxidative phosphorylation via a proton gradient.

Biological "High-Energy" Compounds

Type	Examples
Nucleoside di- and tri-phosphates (anhydrides of phosphoric acid)	ADP, ATP, GTP, etc.
Other anhydrides	1,3-diphosphoglycerate
Enol phosphates	Phosphoenolpyruvate (very energy-rich)
Thioesters	Acetyl CoA; succinyl CoA
Others	Creatine phosphate

V. YIELD OF REDUCED COENZYMES AND ATP FROM THE COMPLETE OXIDATION OF GLUCOSE

Transformation	Enzyme	Reducing Equivalent Produced	~P Produced Substrate Level	Electron Transport Level
Glucose → → 1,3-diphospho- glycerate	Glyceraldehyde- 3-phosphate dehydrogenase	NADH + H$^+$ (x 2)		6 ATP
1,3-Diphospho- glycerate → 3-phosphoglycerate	Phospho- glycerate kinase		(2 ATP)*	
Phosphoenolpyruvate → pyruvate	Pyruvate kinase		2 ATP	
Pyruvate → Acetyl CoA	Pyruvate dehydrogenase	NADH + H$^+$ (x 2)		6 ATP
Isocitrate → α-ketoglutarate	Isocitrate dehydrogenase	NADH + H$^+$ (x 2)		6 ATP
α-Ketoglutarate → succinyl CoA	α-Ketoglutarate dehydrogenase complex	NADH + H$^+$ (x 2)		6 ATP
Succinyl CoA → succinate	Succinyl thiokinase		2 ATP (via GTP)	
Succinate → fumarate	Succinate dehydrogenase	FADH$_2$ (x 2)		4 ATP
Malate → oxaloacetate	Malate dehydrogenase	NADH + H$^+$ (x 2)		6 ATP

* 2ATP's are consumed at start of glycolysis.   4 ATP   34 ATP

Assumptions: (1) tightly-coupled system;
          (2) no energy loss in transferring
NADH from outside of mitochondrion (glycolysis)
to inside. Actually, yield may be reduced, depending on which "shuttle" mechanism is used.

Total 38 ATP from each glucose

VI. REVIEW QUESTIONS ON ENERGETICS AND BIOLOGICAL OXIDATION

DIRECTIONS: Each of the questions or incomplete statements below is followed by five suggested answers or completions. Select the one that is BEST in each case and fill in the corresponding space on the answer sheet.

1. In mitochondrial electron transport, the link between flavoprotein and the cytochrome system is:

A. ferritin
B. NADH - cyt c reductase
C. antimycin A
D. vitamin K
E. CoQ (ubiquinone)

2. How many ATP's are generated in the transfer of electrons from one NADH to oxygen?

A. 1
B. 2
C. 3
D. 4
E. 5

3. Which of the following is NOT correct?

A. A reaction is at equilibrium when the free energy change is zero
B. A reaction may proceed spontaneously if the free energy change is negative.
C. A reaction which would otherwise not proceed can be made to proceed if it is coupled with another reaction for which the free energy change is negative.
D. The free energy change corresponds to (is proportional to) the change in the heat content of the system.
E. The concentration of the reactant molecules are not necessarily equal at equilibrium.

4. A more negative oxidation-reduction system will reduce a more positive system relative to the standard hydrogen electrode. The most positive standard oxidation-reduction potential is shown by which of the following systems?

A. molecular oxygen system
B. NAD system
C. NADP system
D. flavoprotein system
E. cytochrome b system

5. An enzyme

A. increases K_{eq} of the reaction
B. decreases K_{eq} of the reaction
C. increases free energy of reaction
D. decreases free energy of reaction
E. decreases activation energy of reaction

6. Which of the following is NOT involved in mitochondrial electron transport?

A. NADH
B. cytochrome P-450
C. nonheme iron
D. cytochrome b
E. flavoprotein

7. Of the following phosphate compounds, which has a more negative $\Delta G^0{}'$ of hydrolysis than GTP?

A. glucose-1-phosphate
B. glucose-6-phosphate
C. 2,3-diphosphoglycerate
D. phosphoenolpyruvate
E. glycerol-1-phosphate

8. Which of the following is necessary for a reaction to proceed spontaneously?

A. Delta G = 0
B. Delta G < 0
C. Delta G > 0
D. Delta S > 0
E. Delta H < 0

9. When electrons pass from succinate through $FADH_2$ and the electron transport system to oxygen, all of the following are true EXCEPT:

A. P:O ratio is 2
B. Coenzyme Q is involved
C. site I of ATP synthesis is bypassed
D. cytochrome c is involved
E. NADH dehydrogenase is reduced and reoxidized

10. The condensation of oxaloacetate with acetyl CoA has a standard free energy change of -7.7 Kcal/mole. The equilibrium constant for the reaction

A. is greater than one
B. is less than one
C. depends on the concentrations of reactants and products
D. depends upon a catalyst
E. cannot be determined from the data given

11. If the hydrolysis of glucose-6-phosphate has $K_{eq} = 100$, and the phosphorylation of glucose by ATP to form glucose6-phosphate has $K_{eq} = 1000$, then the hydrolysis of ATP to ADP and P_i has $K_{eq} =$

A. 10
B. 1×10^{-1}
C. 1×10^2
D. 1×10^5
E. 1×10^{-5}

12. In a tightly-coupled system, the oxidation reaction producing the most ATP is:

A. isocitrate to malate
B. acetate to CO_2 and H_2O
C. succinate to CO_2 and H_2O
D. succinate to oxalacetate
E. succinate to fumarate

13. The complete metabolism of one mole of pyruvate to CO_2 and H_2O produces about how many moles of ATP?

A. 12
B. 15
C. 30
D. 38
E. 60

14. Creatine phosphate

A. is an intermediate in arginine biosynthesis
B. is a specific inhibitor of aspartate transcarbamylase
C. can serve as an immediate source of energy by direct interaction with actomyosin
D. is an allosteric effector for serine esterase
E. is a "high-energy" compound containing a P-N bond.

15. Catabolism of fat produces more energy per gram than does carbohydrate or protein because:

A. fat yields much acetyl CoA that can enter the Krebs cycle
B. fat-metabolizing tissues produce more heat than carbohydrate-metabolizing tissues
C. fat has a higher (C + H)/O ratio than protein or carbohydrate
D. oxidative metabolism of fat goes most nearly to completion
E. the molecular weight of fatty acids is higher, on average, than that of amino acids or monosaccharides.

16. Which of the following biological processes results in a net increase in the chemical energy of the system?

A. CO_2 fixation
B. anaerobic glycolysis
C. photosynthesis
D. protein synthesis
E. aerobic phosphorylation

17. The most ATP per gram is yielded by which substrate?

A. isocitric acid
B. aspartic acid
C. oleic acid
D. fructose
E. glycogen

18. One reaction can easily be coupled to another if

A. the two reactions have a common intermediate
B. one reaction has a more negative Delta G^0' than the other
C. one reaction has a more positive Delta G^0' than the other
D. the reactions are of the same type
E. one reaction has a greater rate than the other

19. Which of the following is included in oxidative phosphorylation?

A. Phosphoenolpyruvate + ADP $\longrightarrow$ pyruvate + ATP
B. glucose-6-phosphate + ADP $\longrightarrow$ glucose + ATP
C. ADP + phosphate $\longrightarrow$ ATP
D. UTP + ADP $\longrightarrow$ UDP + ATP
E. 2ADP $\longrightarrow$ ATP + AMP

20. Uncoupling of mitochondrial oxidative phosphorylation results in

A. continued ADP formation, halt in O2 consumption
B. slowing down of the Krebs cycle
C. inhibition of mitochondrial membrane ATPase
D. halt in ATP formation but continued O_2 consumption
E. halt in mitochondrial metabolism

21. Dinitrophenol, an uncoupler of oxidative phosphorylation

A. inhibits NAD^+ - requiring reactions
B. inhibits cytochromes
C. inhibits respiration without affecting ATP synthesis
D. permits electron transport without ATP synthesis
E. inhibits ATP synthesis and respiration

22. The equilibrium constant for the reaction:

malate + NAD^+ $\rightleftharpoons$ oxalacetate + NADH + H^+

is defined by

A. $K = \dfrac{(oxalacetate)(NAD^+)(H^+)}{(malate)(NADH)}$

B. $\dfrac{1}{K} = \dfrac{(oxalacetate)(NADH)(H^+)}{(malate)(NAD^+)}$

C. $K = \dfrac{(oxalacetate)(NADH)(H^+)}{(malate)(NAD^+)}$

D. $K = \dfrac{(malate)(NAD^+)}{(oxalacetate)(NADH)(H^+)}$

E. $K = \dfrac{(malate)(NAD^+)(H^+)}{(oxalacetate)(NADH)}$

23. How many high-energy phosphates can result from the over-all reaction:

isocitrate + 3/2 O_2 $\rightleftharpoons$ fumarate + $2CO_2$ + $2H_2O$?

A. 4
B. 6
C. 7
D. 9
E. 10

24. Which of the following is always involved in biological oxidation-reduction reactions?

A. transfer of hydrogens
B. formation of water
C. mitochondria
D. transfer of electrons
E. direct participation of oxygen

25. If a preparation of healthy mitochondria is incubated with excess succinate, which of the following will stimulate oxygen uptake the most?

A. glucose-1-phosphate
B. oligomycin
C. malonate
D. ATP
E. ADP + P_i

26. About how much tristearin will yield the same amount of energy as that obtained from the metabolism of 100 grams of glucose?

A. 20
B. 40
C. 80
D. 100
E. 180

27. All of the following are "high-energy" compounds except:

A. phosphocreatine
B. phosphoenolpyruvate
C. AMP
D. ADP
E. ATP

28. When NADH is transformed to NAD^+, it loses

A. one hydronium ion
B. one hydride ion
C. one electron
D. two electrons
E. two protons and one electron

29. Cytochrome oxidase contains

A. Cobalt
B. Zinc
C. Magnesium
D. Vanadium
E. Copper

30. Mitochondria

A. carry out glycolysis
B. make lipoprotein
C. conduct oxidative phosphorylation
D. synthesize sterols
E. hydroxylate drugs

31. Cytochrome oxidase reacts specifically with

A. Carbon dioxide
B. periodic acid – Schiff reagent
C. parachloromercuribenzoate
D. cyanide
E. diisopropylfluorophosphate

32. The complete oxidation of one mole of glucose in a biological system leads to the formation of about how many moles of ATP from ADP and P_i?

A. 2
B. 7
C. 14
D. 35
E. 70

DIRECTIONS: The group of items below consists of lettered headings followed by a set of numbered words or phrases. For each numbered word or phrase, select the ONE lettered heading that is most closely associated with it and fill in the corresponding space on the answer sheet. Each heading may be used once, more than once, or not at all.

A. Inner membrane of mitochondria
B. Outer membrane of mitochondria
C. Space between the two membranes
D. Matrix
E. Cytoplasm

33. Pyruvate dehydrogenase complex

34. Cytochrome respiratory chain

35. Glycolytic pathway

DIRECTIONS: For each of the questions or incomplete statements below, ONE or MORE of the answers or completions is correct. On the answer sheet fill in space

A if only 1, 2, and 3 are correct
B if only 1 and 3 are correct
C if only 2 and 4 are correct
D if only 4 is correct
E if all are correct

FILL IN ONLY ONE SPACE ON YOUR ANSWER SHEET FOR EACH QUESTION

Directions Summarized				
(A) 1,2,3 only	(B) 1,3 only	(C) 2,4 only	(D) 4 only	(E) All are correct

36. In the reaction below, which of the following is true?

COOH   phosphoglyceromutase   COOH
| ⇌ |
HCOH   Delta $G^{0'}$ = +1.06 Kcal   HC-OPO$_3$$^-$
| at 25°C |
CH$_2$OPO$_3$$^-$ CH$_2$OH

1. The K_m can be calculated from the amounts of substrates at equilibrium.
2. The reaction can occur in both aerobic and anaerobic metabolism.
3. The reaction goes faster at 20°C than at 25°C.
4. At equilibrium the substrate on the left has a higher concentration than the one on the right.

37. Which of the following substrates is able to phosphorylate ADP to ATP?

1. 3-phosphoglyceraldehyde
2. 1,3-diphosphoglycerate
3. fructose-1,6-diphosphate
4. phosphoenolpyruvate

38. Concerning entropy, which of the following is (are) correct?

1. Entropy means the degree of disorder or randomness.
2. Living things maintain low entropy by producing an increase in entropy of the surroundings.
3. Denatured macromolecules have higher entropy than the native forms.
4. A reaction may be spontaneous even if its change in entropy is negative.

39. Which of the following glycolytic reactions can phosphorylate ADP?

1. hexokinase
2. pyruvate kinase
3. phosphofructokinase
4. 3-phosphoglycerate kinase

40. Which of the following can regulate the generation of ATP from glucose in muscle?

1. ATP
2. ADP
3. P_i
4. O_2

41. In catalysis of a reaction, an enzyme changes the

1. rate
2. Delta G
3. activation energy
4. K_{eq}

42. In mitochondrial electron transfer from cytochrome b to cytochrome c

1. cytochrome c Fe^{+++} is reduced to cytochrome c Fe^{++}
2. it occurs because cyt b is a stronger oxidizing agent than cyt c
3. enough energy is produced for phosphorylation of one ADP to ATP
4. cytochrome b Fe^{+++} is oxidized to cytochrome b Fe^{++}

FILL IN ONLY ONE SPACE ON YOUR ANSWER SHEET FOR EACH QUESTION

Directions Summarized				
(A) 1,2,3 only	(B) 1,3 only	(C) 2,4 only	(D) 4 only	(E) All are correct

43. In an enzymatic reaction, Delta $G^{o'}$ is

1. equal to the heat produced
2. equal to the activation energy
3. proportional to the concentration of enzyme
4. proportional to $-\log K_{eq}$

44. What is the consequence of the oxidation of one mole of acetyl CoA via the TCA cycle?

1. net consumption of 2 O_2
2. net production of 2 ATP
3. net production of 2 CO_2
4. net consumption of 1 oxalacetate

DIRECTIONS: The set of lettered headings below is followed by a list of numbered words or phrases. On the answer sheet, for each numbered word or phrase fill in space

        A if the item is associated with (A) only
        B if the item is associated with (B) only
        C if the item is associated with both (A) and (B)
        D if the item is associated with neither (A) nor (B)

Questions 45 - 47:

    A. ATP is used
    B. ATP is produced
    C. Both
    D. Neither

45. Glycogen ⟶ glucose

46. Glucose ⟶ glycogen

47. Glucose-1-phosphate ⟶
       3-phosphoglyceric acid

VII. ANSWERS TO QUESTIONS ON
ENERGETICS AND BIOLOGICAL OXIDATION

1. E		25. E	
2. C		26. B	
3. D		27. C	
4. A		28. B	
5. E		29. E	
6. B		30. C	
7. D		31. D	
8. B		32. D	
9. E		33. D	
10. A		34. A	
11. D		35. E	
12. C		36. C	
13. B		37. C	
14. E		38. E	
15. C		39. C	
16. C		40. E	
17. C		41. B	
18. A		42. B	
19. C		43. D	
20. D		44. B	
21. D		45. D	
22. C		46. A	
23. D		47. C	
24. D			

4. CARBOHYDRATES

Robert E. Hurst

I. INTRODUCTION

Carbohydrate metabolism is the core of intermediary metabolism, providing a large part of the energy requirements of the organism, short-term storage of energy in the form of glycogen, and carbon skeletons for biosynthesis. Amino acids and lipids feed into the pathways of carbohydrate metabolism and the citric (tricarboxylic) acid cycle. The following material is a concentrated essence of carbohydrate metabolism organized to show the pathways and their regulation. Important structures are given at the end of the chapter.

II. DIGESTION

Carbohydrates are present in the diet as simple sugars, disaccharides, and polysaccharides. **Starch**, the major dietary polysaccharide, is cleaved by salivary and pancreatic amylases to the disaccharide **maltose** and oligosaccharides which contain α-1,6 linkages. Disaccharides are cleaved in the small intestine by disaccharidase enzymes of varying specificities to monosaccharides. Examples include maltase, which cleaves maltose to **glucose**, and lactase, which cleaves lactose to glucose and **galactose**. The resulting monosaccharides are then absorbed through the intestine into the bloodstream.

While infants of all races possess sufficient lactase in the intestine, adults of many races lack lactase, thereby giving rise to lactose intolerance. Intolerant individuals who consume more than a few ounces of milk will experience diarrhea and intestinal gas caused by microbial fermentation of lactose in the gut. Lactase persists in adults whose ancestors came from parts of the world with a long history of consuming milk, such as northern Europe, parts of central Asia, and parts of Africa. Most Blacks and Orientals are lactose-intolerant, as are significant proportions of Semitic and Mediterranean Caucasians.

III. METABOLISM OF MONOSACCHARIDES

A. Overview

Glucose is metabolized directly, while other sugars such as galactose and mannose are converted into products which can be metabolized along the glucose pathways.

The first step in the metabolism of glucose is its entry into the cell. Glucose passes through the cell membrane by an active transport system. **Insulin** is also required for the entry of glucose into cells.

Immediately upon entering the cell, glucose is phosphorylated by the enzyme <u>hexokinase</u> to **glucose 6-phosphate (G6P)**:

$$\text{Glucose} + \text{ATP} \xrightarrow{\text{hexokinase}} \text{G-6-P} + \text{ADP}$$

This step is essentially irreversible. Brain and muscle tissue, among others do not contain <u>glucose 6-phosphatase</u>, the enzyme which reverses this step. Thus glucose enters these cells but does not leave them. Instead of hexokinase, liver parenchymal cells contain <u>glucokinase</u>, which has markedly different kinetic properties. In addition to not being inhibited by its product (which prevents the cell from sequestering all its phosphate as phosphohexoses), the K_m of glucokinase is about 10 mM instead of the 0.1 mM of hexokinase. At the usual blood glucose concentration of 5 mM, this difference contributes markedly to the buffering of blood glucose. Increases of blood glucose increases its uptake by the liver, where it is converted to glycogen, while decreases cause the converse to occur.

G-6-P can then be metabolized by one of several pathways, as summarized below. Each pathway also has a stoichiometry which, in general, summarizes its function. The reactions shown below do not include water, inorganic phosphate, or protons and are, therefore, not balanced equations. Glycolysis serves to convert G-6-P to pyruvate and ATP.

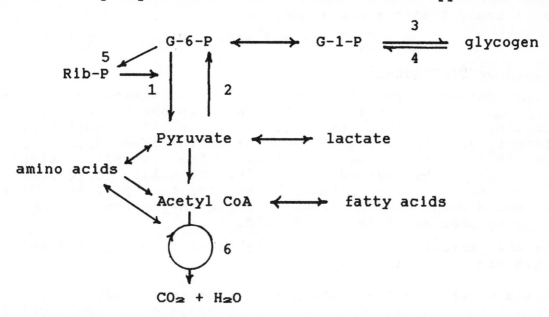

B. Pathways and Stoichiometry

1. **Glycolysis** (metabolizes glucose to 3-carbon unit plus ATP and NADH, which is convertible to 3 ATP in the electron transport system):

$$\text{Glucose} + 2\ \text{ADP} + 2\ \text{NAD}^+ \longrightarrow 2\ \text{pyruvate} + 2\ \text{ATP} + 2\ \text{NADH}$$

2. **Gluconeogenesis** (Resynthesizes glucose from pyruvate. Useful in processing excess lactate produced by muscle and for converting certain amino acids to glucose):

$$2 \text{ Pyruvate} + 4 \text{ ATP} + 2 \text{ GTP} + 2 \text{ NADH} \longrightarrow$$
$$\text{Glucose} + 4 \text{ ADP} + 2 \text{ GDP} + 2 \text{ NAD}^+$$

3. **Glycogen Synthesis** (Converts glucose to a storage form, glycogen, capable of meeting about 24 hours of carbohydrate metabolism requirements):

$$\text{glycogen}_n + \text{UDP-glucose} \longrightarrow \text{glycogen}_{n+1} + \text{UDP}$$

4. **Glycogenolysis** (Converts the storage form, glycogen, to the metabolizable form, G-1-P):

$$\text{glycogen}_{n+1} + P_i \longrightarrow \text{glycogen}_n + \text{G-1-P}$$

5. **Pentose Phosphate Pathway** (Synthesizes 5-carbon sugars and NADPH, which is used for reductive biosynthesis, particularly in lipid pathways):

$$\text{G-6-P} + 2 \text{ NADP}^+ \longrightarrow CO_2 + 2 \text{ NADPH} + \text{ribulose 5-P}$$

6. **Citric Acid Cycle** (Converts Acetyl CoA to carbon dioxide + water + energy through the electron transport system in a cyclic series of reactions):

$$\text{Acetyl CoA} + 3 \text{ NAD}^+ + \text{FAD} + \text{GDP} \longrightarrow$$
$$2 CO_2 + 3 \text{ NADH} + FADH_2 + \text{GTP} + \text{CoA}$$

C. Glycolysis-Gluconeogenesis

These two pathways operate reciprocally and, therefore, are regulated in a manner that, in a given tissue, only one is operating at a particular time. One is not the reverse of the other; note that they have different stoichiometries. Several of the same enzymes operate in a reversible manner in each pathway. The reversible steps are shown as single, double-headed arrows (Figure 4.1). There are three irreversible steps in each pathway where the two pathways differ. These are shown with double arrows and two enzyme names. Important regulatory compounds are noted at the side.

Enzymes and Reaction Types (in parentheses) in Glycolysis and Gluconeogenesis are listed in Table 4.1.

Glycolysis occurs in the cytoplasm. The end product, pyruvate, crosses the mitochondrial membrane. Gluconeogenesis can occur from either pyruvate or oxaloacetate. In starting from pyruvate, the first of the two steps in gluconeogenesis which reverse the strongly exergonic pyruvate kinase reaction at the cost of 2 ATP, occurs in the mitochondrion. The product of the two reactions, phosphoenolpyruvate, can cross back into the cytosol. Alternately, gluconeogenesis can start with oxaloacetate, which can be provided either by amino acid metabolism or by the citric acid cycle. In the latter case, oxaloacetate, which cannot traverse the mitochondrial membrane directly, can indirectly traverse it by being transaminated to aspartate, which can traverse the membrane. Once in the cytoplasm, aspartate is converted

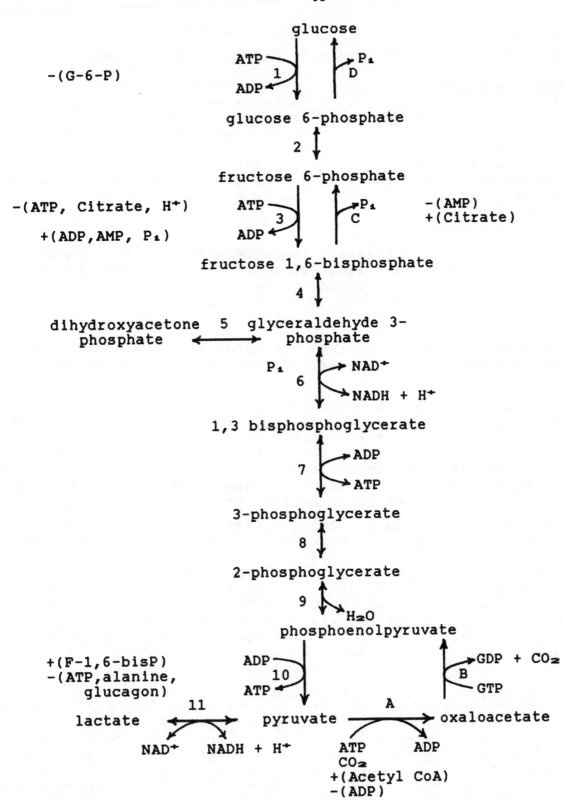

Figure 4.1. Glycolysis-Gluconeogenesis.

<u>Glycolysis</u>

1. **Hexokinase** or Glucokinase (a)
2. Phosphohexose isomerase (c)
3. <u>**Phosphofructokinase**</u> (a)
4. Aldolase (e)
5. Phosphotriose isomerase (c)
6. Glyceraldehyde 3-phosphate dehydrogenase (f)
7. Phosphoglycerate kinase (a)
8. Phosphoglycerate mutase (b)
9. Enolase (d)
10. **Pyruvate kinase** (a)
11. Lactate dehydrogenase (g)

<u>Gluconeogenesis</u>

A. <u>**Pyruvate Carboxylase**</u> (h)
B. Phosphoenolpyruvate carboxykinase (i)
C. **Fructose 1,6-bis phosphatase** (j)
D. Glucose 6-phosphatase (j)

*Important regulatory enzymes are shown in bold type; the main one is also underlined.

<u>Table 4.1</u>. a = phosphoryl transfer (phosphorylation), b = phosphoryl shift, c = isomerization, d = dehydration, e = aldol cleavage/condensation, f = phosphorylation coupled to oxidation, g = reduction, h = carboxylation, i = decarboxylation, j = hydrolysis

back to oxaloacetate, which is then converted to phosphoenolpyruvate by phosphoenolpyruvate carboxykinase. This enzyme is found in both the cytosol and mitosol.

Gluconeogenesis provides a means for other metabolites, particularly amino acids, to be converted into glucose. This allows for efficient use of foodstuffs and for maintenance of blood glucose levels long after all dietary glucose has been metabolized. Only leucine and lysine are not capable of contributing carbon atoms to net glucose synthesis. Fatty acids are not capable of contributing to gluconeogenesis. The glycerol moiety of triglycerides can, however.

These pathways form cycles involving different organs. Much of the energy for muscle contraction is provided by ATP generated during glycolysis. The process generates high concentrations of pyruvate and lactate. The latter is exported to the blood which carries it to the liver and kidney where gluconeogenesis converts it back to glucose. This process is referred to as the **Cori cycle**, which also operates in erythrocytes. In the **alanine cycle**, pyruvate in muscle is transaminated to alanine, which is then transported by blood to the kidney where it is deaminated. The amino group is converted to urea, while the pyruvate is converted back to glucose by gluconeogenesis.

These pathways are carefully regulated so that one is off while the other is on, and so that glycolysis is active when AMP and fructose 1,6-bisphosphate are high. In general, most pathways are regulated at the <u>committed step</u>, which is the first irreversible, completely unique step in the pathway. Since G-6-P has several possible fates other than glycolysis, it is not a unique step; the phosphorylation of F-6-P is, however. In general, the enzymes of the regulated steps are **allosteric** and respond to allosteric effectors; indeed, metabolism can be thought of as being a chemical computer, with the concentrations of a few key metabolites providing signals which turn systems on and off.

The synthesis of these key enzymes are often subject to **induction** and **repression** as well.

Three end products of glycolysis, protons, ATP and citrate, an eventual product of glycolysis in the citric acid cycle, signal a halt to glycolysis by serving as allosteric effectors. High concentrations of glucose 6-phosphate inhibit gluconeogenesis. The main signal for gluconeogenesis is a high concentration of Acetyl CoA, which will be present when lipid metabolism is satisfying the need for this key metabolite. Thus, although fatty acids cannot be converted to glucose through gluconeogenesis, gluconeogenesis is facilitated by a high level of fatty acid metabolism. Pyruvate kinase is also inactivated by covalent modification (phosphorylation by a cAMP-dependent protein kinase-- see glycogen metabolism for explanation). Pyruvate kinase and glucokinase are also induced by high levels of blood glucose and insulin.

Inherited diseases of the glycolytic pathway are rare, most probably being lethal. Deficiency of erythrocyte pyruvate kinase is known and results in a severe hemolytic anemia. Because they lack mitochondria, erythrocytes are absolutely dependent upon glycolysis for energy.

D. Entry of Other Sugars

Other sugars are metabolized through the above pathways after conversion to a glycolytic intermediate.

Glycerol:

glycerol $\xrightarrow{\quad 1 \quad}$ glycerol 3-phosphate $\xrightarrow{\quad 2 \quad}$ dihydroxyacetone phosphate

      ATP   ADP   NAD$^+$ NADH + H$^+$

1. glycerol kinase (phosphoryl transfer)
2. glycerol phosphate dehydrogenase (oxidation)

Glycerol 3-phosphate is also used for fat biosynthesis.

Fructose:

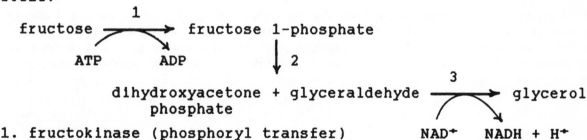

1. fructokinase (phosphoryl transfer)
2. Aldolase B (aldol condensation)
3. Alcohol dehydrogenase (oxidation)

Galactose:

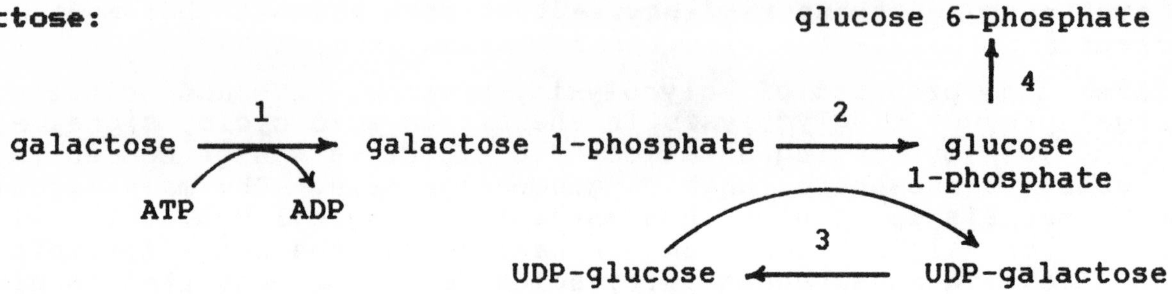

glucose 6-phosphate

galactose $\xrightarrow{\text{1}}$ galactose 1-phosphate $\xrightarrow{\text{2}}$ glucose 1-phosphate

ATP ADP

UDP-glucose $\xleftarrow{\text{3}}$ UDP-galactose

4

1. Galactokinase (phosphoryl transfer)
2. Galactose 1-phosphate uridyltransferase (group transfer)
3. UDP-galactose 4-epimerase (epimerization)
4. Phosphoglucomutase (phosphoryl shift)

In the second reaction the UDP group is transferred from glucose to galactose, and the glucose 1-phosphate is converted to the glycolytic intermediate, glucose 6-phosphate. The UDP-glucose consumed by the second reaction is regenerated by the third reaction, leading to a cycle.

Essential fructosuria, a benign disorder, results from a lack of hepatic fructokinase. Deficiency of aldolase B results in **hereditary fructose intolerance,** a disease characterized by fructose-induced hypoglycemia and liver damage. In the presence of fructose, the resulting high level of fructose 1-phosphate inhibits liver phosphorylase, stopping glucose production from glycogen (see glycogen metabolism). Tying up all the cell's phosphate virtually stops ATP synthesis in the liver, which apparently prevents the maintenance of normal ionic gradients and leads to osmotic damage to hepatocytes. **Galactosemia** results from a deficiency of either galactokinase, which produces a relatively mild disease, or the uridyltransferase. In the latter case, the disease is more severe because of accumulation of galactose 1-phosphate.

IV. PENTOSE PHOSPHATE PATHWAY

This pathway consists of two branches, an oxidative branch as shown below, and a non-oxidative branch. In the latter, the product of the oxidative branch, ribulose 5-phosphate is converted into a series of other pentoses, a tetrose, and a heptose. The pathway connects back to glycolysis at fructose 6-phosphate and at glyceraldehyde 3-phosphate. Interestingly, there are no irreversible steps between glycolysis and the non-oxidative branch which therefore freely interconnect, making it possible to synthesize ribose and other sugars even when the oxidative pathway is shut off.

The main function of the oxidative branch appears to be the production of **NADPH,** which is needed for reductive biosynthesis, since NADPH and NADH metabolism are completely independent. It is particularly active in adipose tissue and is low in muscle tissue. The first step involves oxidation of the aldehyde on C-1 to a carboxyl group, proceeding through an intermediate lactone (not shown). The second step involves decarboxylation with the formation of a ketose.

$$\text{Glucose 6-phosphate} \xrightarrow{1} \text{6-phosphogluconate} \xrightarrow{2} \text{ribulose 5-phosphate} + CO_2$$

$$\text{NADP}^+ \quad \text{NADPH} + H^+ \qquad\qquad \text{NADP}^+ \quad \text{NADPH} + H^+$$

1. Glucose 6-phosphate dehydrogenase
2. 6-Phosphogluconate dehydrogenase

The first enzyme represents the control point. It is inhibited by NADPH and ATP, which couples the availability of both substrates for reductive biosynthesis to the production of NADPH

The non-oxidative branch can be summarized as described below.

$$\text{Ribulose 5-P} \xrightarrow{1} \text{ribose 5-P}$$
$$\xrightarrow{2,3,4} \text{F-6-P + glyceraldehyde 3-P}$$

1. Phosphopentose isomerase
2. Phosphopentose epimerase
3. Transketolase (requires thiamine pyrophosphate cofactor)
4. Transaldolase

This pathway is unique in that it can function in distinct modes, depending upon the relative needs for its products (NADPH and pentose) and upon the status of other pathways.

<u>Mode 1</u>: The need for NADPH is much greater than for ribose, and the cell is in an anabolic phase. Glucose is metabolized to CO_2 and NADPH.

Two moles of NADPH and one mole of pentose are produced by the oxidative branch. Pentose is metabolized by the nonoxidative branch to fructose 6-phosphate and glyceraldehyde 3-phosphate. Glucose 6-phosphate is resynthesized from these substrates and is recycled through the oxidative branch.

$$\text{G-6-P} + 12 \text{ NADP}^+ + 7 \text{ H}_2\text{O} \longrightarrow 6 \text{ CO}_2 + 12 \text{ NADPH} + 12 \text{ H}^+ + P_i$$

<u>Mode 2</u>: The need for NADPH is much greater than for ribose, and the cell is in a catabolic phase. Glucose is metabolized to NADPH and pyruvate.

Fructose 6-phosphate and glyceraldehyde 3-phosphate derived from the oxidative pathway are metabolized by glycolysis to pyruvate.

$$3 \text{ G-6-P} + 6 \text{ NADP}^+ + 5 \text{ NAD}^+ + 5 \text{ P}_i + 8 \text{ ADP} \longrightarrow$$
$$5 \text{ pyruvate} + 3 \text{ CO}_2 + 6 \text{ NADPH} + 5 \text{ NADH} + 8 \text{ ATP} + 2 \text{ H}_2\text{O} + 8 \text{ H}^+$$

<u>Mode 3</u>: The need for NADPH and ribose are balanced. Glucose is metabolized to ribose.

G-6-P is metabolized through the oxidative branch to ribose 5-P, which is then utilized for nucleotide biosynthesis.

$$G-6-P + 2NADP^+ + H_2O \longrightarrow \text{ribose } 5-P + 2 \text{ NADPH} + 2 \text{ H}^+ + CO_2$$

<u>Mode 4</u>: Much more ribose 5-P is needed than NADPH. The oxidative branch is inactive.

Fructose 6-P and glyceraldehyde 3-P from the glycolytic pathway are metabolized by the nonoxidative branch to yield ribose. The lack of irreversible steps between glycolysis and the nonoxidative branch makes this series of reactions possible. The oxidative branch is turned off by high NADPH concentrations.

$$5 \text{ G-6-P} + ATP \longrightarrow 6 \text{ ribose } 5-P + ADP + H^+$$

The most important disease involving this pathway is <u>glucose 6-phosphate dehydrogenase (G6PD) deficiency</u>. The most striking symptom is **hemolytic anemia**, which is frequently triggered by certain drugs such as the antimalarial **pamaquine**. Apparently, NADPH is needed to reduce glutathione in erythrocyte membranes, which functions to prevent oxidation of lipids and other substances in the membrane. The build-up of oxidized substances appears to destabilize the membrane, leading to hemolysis. The gene for this disorder is much more common among races which are indigenous to areas with a high incidence of malaria.

V. GLYCOGENOLYSIS-GLYCOGENESIS

These two pathways control the synthesis and degradation of glycogen, which is the storage form of carbohydrate metabolism. Glycogen storage is concentrated in the liver and skeletal muscle. The amount of glucose in the body fluids of a 70-kg man is equivalent to about 40 kcal, while the total body glycogen is equivalent to at least 600 kcal, even after an overnight fast.

Glycogen is composed of glucose units linked α-1,4 with about 1 in 10 residues being linked α-1,6. The latter lead to a highly branched structure. Since the non-reducing termini are the metabolically most active parts of the molecule, the branched structure increases the rate at which glucose can be added to and removed from the glycogen molecule.

A. <u>Glycogenolysis</u>

Glycogen is degraded in two distinct steps. The first step consists of the stepwise removal of α-1,4 groups from the non-reducing termini (to the left in the figure) by the enzyme <u>glycogen phosphorylase</u>. This reaction is reversible under ordinary conditions, but it is not the mechanism used for glycogen synthesis. Note that the product of this reaction is glucose 1-phosphate, which is readily con-

verted to glucose 6-phosphate, a glycolytic intermediate. Phosphorylase will degrade glycogen until the stub of the branch is 4 glucose
units in length. At this point, a transferase enzyme moves the "stub"
to the non-reducing terminus of another "stub", yielding a piece which
can be further cleaved by phosphorylase and yielding a single glucose
attached α-1-6. An α-1,6-glucosidase, or "debranching enzyme" cleaves
this glucose, releasing glucose instead of G-1-P.

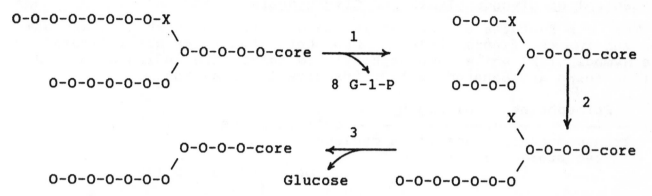

1. Phosphorylase
2. Transferase
3. Debranching enzyme

The final product shown above is then further degraded by phosphorylase to G-1-P. Note that this pathway is energetically favorable,
since the product, G-1-P, is phosphorylated and can be converted to G-
6-P by phosphoglucomutase without using ATP.

B. Glycogenesis

The first step of glycogen synthesis consists of the formation of
uridine diphosphoglucose (UDPG), a nucleotide sugar. Such nucleotide
sugars are the biosynthetically active forms of sugars. This reaction
is freely reversible *in vitro*. *In vivo*, however, because of the presence of very active inorganic pyrophosphatases, the pyrophosphate is
rapidly cleaved to two moles of inorganic phosphate, driving the reaction towards the right. The enzyme glycogen synthetase adds glucose
units to the non-reducing terminus using UDPG as its substrate.
Branch points are introduced by a branching enzyme which cleaves a
string of α-1,4-residues and moves the segment containing the non-reducing terminus to a more interior site where it is attached to the 6-
position of a glucose molecule.

$$G\text{-}1\text{-}P + UTP \xrightleftharpoons{\ 1\ } UDPG + PP_i \xrightarrow{\ 2\ } 2P_i$$

$$UDPG + glycogen_n \xrightarrow{\ 3\ } glycogen_{n+1} + UDP$$

1. UDP-glucose pyrophosphorylase
2. pyrophosphatase
3. glycogen synthetase

The energy conservation from the breakdown of glycogen is 90 % efficient, since only about 1 residue in 10 is released in the form of free glucose which must be re-phosphorylated before it can be metabolized. Since the metabolism of 1 mole of G-6-P yields 37 moles of ATP but requires the expenditure of only 1 mole of ATP, the recovery of ATP-energy is over 97 %.

C. Control of Glycogenolysis and Glycogenesis

The main features of the regulation of glycogen metabolism are summarized in the figure below. Phosphorylated forms are indicated by the symbol (+P) while dephosphorylated forms are indicated by (-P). Active forms are denoted as "a", inactive forms as "b".

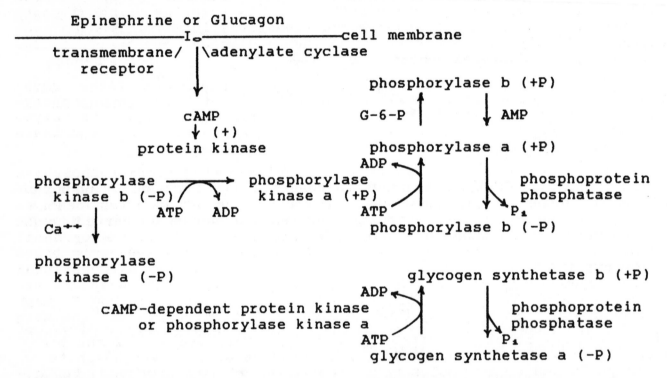

Glycogen metabolism is regulated from outside the cell by the hormones **glucagon**, which is active primarily in liver, and **epinephrine**, which is active primarily in muscle. Glucagon is released by the α-cells of the pancreas, while epinephrine is secreted by the neuromuscular plate in response to neural stimulation and by the adrenal medulla. The end result of the binding of these proteins to the outside of cells is the phosphorylation of both phosphorylase and glycogen synthetase within the cell. The results of these phosphorylations are quite different for the two enzymes. Phosphorylation inactivates glycogen synthetase while it activates phosphorylase. *Thus, glucagon or epinephrine will turn on glycogen degradation and stop glycogen synthesis.* This form of regulation is referred to as regulation by covalent modification.

In modern nomenclature, regardless of whether phosphorylation activates or deactivates the enzyme, *the active form will always be called*

the "a" form and the inactive form will be called the "b" form.
MNEMONIC: "a equals active".

1. Glucagon or epinephrine binds to receptors on the cell surface. These receptors form part of a transmembrane protein complex,
adenylate cyclase, which synthesizes cyclic AMP (cAMP) on the inner
face of the membrane, releasing cAMP into the cytoplasm. *Thus, the
presence of the hormones on the outside of the cell causes synthesis
of cAMP within the cell.*

2. The cAMP activates a protein kinase (cAMP-dependent protein
kinase) in the cytosol. This enzyme is inactive in the absence of
cAMP. In turn, protein kinase phosphorylates both glycogen synthetase
and phosphorylase kinase on specific serine residues. The dephospho
(active) form of the glycogen synthetase is referred to as glycogen
synthetase a while the phosphorylated (inactive) form is glycogen synthetase b. *Glycogen synthetase, and hence glycogen synthesis, is inhibited by phosphorylation.*

3. Phosphorylase kinase is activated by phosphorylation. Phosphorylase kinase a (phosphorylated) phosphorylates glycogen phosphorylase b, converting it to the active form phosphorylase a, the enzyme
mainly responsible for glycogen degradation. *Phosphorylase, and hence
glycogen degradation, is stimulated by phosphorylation.*

4. Glycogen metabolism is also regulated by glucose 6-phosphate
which is a strong inhibitor of phosphorylase a and an activator of
glycogen synthetase b. Previously, glycogen synthetase b was known as
glycogen synthetase D because its activity was dependent upon glucose
6-phosphate, while the "a"-form was known as the "I" (or independent)
form. Thus, glycogen degradation can be halted and synthesis begun
independent of glucagon or epinephrine levels in the extracellular
fluid whenever levels of the key metabolite glucose 6-phosphate are
high and signal that the cell has sufficient or excess glucose to meet
its metabolic needs. This mechanism is considered to be the more
"primitive" regulatory mechanism and covalent modification is generally considered to be the more significant.

5. Phosphorylase kinase b in muscle is also subject to allosteric regulation by calcium ions, which activate the normally inactive dephospho form. When activated by calcium ions, the normally inactive phosphorylase b (which is now called phosphorylase kinase a,
even though it is, in fact, different from the "a" form produced by
phosphorylation) is capable of phosphorylating both phosphorylase and
glycogen synthetase. Since release of calcium ions is one of the
first steps in muscle contraction, this provides a means by which
glycogen degradation can be initiated in muscle virtually instanta-
neously, even before the epinephrine released by the motor neurons can
effectively activate the system through cAMP. Phosphorylase kinase b
consists of four, non-identical subunits. One of these, the calmodulin subunit, binds calcium ions tightly and induces a conformational
change in the rest of the subunits which activates the normally inactive phosphorylase kinase b. Other proteins subject to calcium regulation also contain a calmodulin subunit.

6. The effects of phosphorylation may be reversed by <u>phosphodiesterase</u>, an enzyme which destroys cAMP by converting it to AMP, and by <u>phosphoprotein phosphatase</u>, which hydrolyzes the phosphate groups added to the protein subunits in response to rising cAMP levels. Thus, if epinephrine or glucagon levels are low in the extracellular fluid, cAMP levels will fall in response to phosphodiesterase activity and the lack of new synthesis. In turn, this will inactivate cAMP-dependent kinase, and phosphoprotein phosphatase will convert the phosphorylated forms of the enzymes to the dephospho- forms. Note that this will activate glycogen synthetase but will deactivate glycogen phosphorylase.

7. The activity of phosphoprotein phosphatase is also regulated. The cAMP-dependent kinase also phosphorylates a protein called the **inhibitory protein**, or I protein. When phosphorylated (on threonine in this case), the a-form of I-protein is a potent inhibitor of phosphoprotein phosphatase activity against phosphorylase kinase a, phosphorylase a, and glycogen synthetase b, but not against dephosphorylation of I-protein a. This prevents futile cycling while still allowing the system to respond to hormonal stimulation since falling levels of cAMP will first be manifested by dephosphorylation of I-protein, then by dephosphorylation of the other enzymes. The regulation of phosphoprotein phosphatase is not shown.

8. To summarize, the enzymes in the glycogen metabolic pathways exist in either of two states; a phosphorylated or a dephosphorylated state. In the presence of epinephrine or glucagon, most of the enzymes are phosphorylated; in the absence of the hormones, most of the enzymes are not phosphorylated. In the phosphorylated state, glycogen degradation is "on" while glycogen synthesis is "off". In the dephosphorylated state, glycogen synthesis is "on" while degradation is "off".

D. <u>Diseases of Glycogen Metabolism</u>

A number of diseases of glycogen metabolism have been identified. Depending upon which enzyme is deficient, these diseases can manifest either storage of excessive levels of glycogen or the synthesis of glycogen of abnormal structure, or both. Their characteristics are summarized in Tables 4.2 and 4.3.

VI. THE CITRIC ACID CYCLE

This is the final pathway in the metabolism of carbohydrates, amino acids and lipids and results in the conversion of acetyl CoA to carbon dioxide, water, and energy in the form of ATP and NADH. The reaction sequence is cyclic; that is one of the starting reactants (oxaloacetate) is regenerated. The **Citric Acid Cycle** is also known as the **tricarboxylic acid cycle** or the **Krebs Cycle**. The enzymes for the Citric Acid Cycle are located in the mitochondrion.

Type or name	Defective enzyme	Organ affected	Glycogen levels and structure*
I. von Gierke's Disease	Glucose 6-phosphatase	Liver & Kidney	I, N
II Pompe's Disease	α-1,4-glucosidase	All organs	very I, N
III Cori's Disease	debranching enzyme	Muscle and Liver	I, A: short outer branches
IV Andersen's	branching enzyme	Liver and Spleen	I, A: v. long outer branches
V McArdle's Disease	phosphorylase	Muscle	I, N
VI Hers' Disease	phosphorylase	Liver	I, N
VII	phosphofructokinase	Muscle	I, N
VIII	phosphorylase kinase	Liver	I, N

*I = increased, N = normal levels or structure, A = abnormal structure

Table 4.2. Glycogen Storage Diseases: Nomenclature and Effects on Glycogen

A. Pyruvate Dehydrogenase Complex

While the end product of lipid metabolism is acetyl CoA produced in the mitochondrion, the end product of carbohydrate metabolism is pyruvate produced in the cytosol. Pyruvate must first be transported across the mitochondrial membrane and then converted to acetyl CoA. This transformation is carried out by a multienzyme complex, the pyruvate dehydrogenase complex, located in mitochondria. The reaction is essentially irreversible. It is the irreversibility of this step which prevents fatty acids from being converted to glucose. The overall reaction is:

$$\text{pyruvate} + NAD^+ + \text{CoASH} \xrightarrow{\text{pyruvate dehydrogenase}} \text{Acetyl CoA} + NADH + H^+ + CO_2$$

Type	Clinical Features	Mechanisms
I	hepatomegaly, severe hypoglycemia, ketosis, hyperlipidemia, acidemia	gluconeogenesis stops at G-6-P, causing hypoglycemia, high lactate and acetyl CoA levels and blocking glycogen degradation. High levels of acetyl CoA result in overactive lipid metabolism.
II	cardiorespiratory failure, usually before age 2.	Function of this enzyme is not entirely clear, since glycogen metabolism is apparently normal. No metabolic consequences.
III	similar to type I but milder.	Less severe because phosphorylase is able to degrade glycogen to some extent, limiting degree of hypoglycemia.
IV	progressive cirrhosis of liver, liver failure usually before age 2	Accumulation is apparently toxic to liver as well as to heart.
V	painful muscle cramps on exercising, limited endurance	Muscle has no energy stores due to unavailability of glycogen. No other metabolic consequences.
VI	similar to Type I, but milder	Gluconeogenesis is only source of blood glucose, causing a mild hypoglycemia
VII	similar to Type V	Block in muscle glycolysis means no muscular endurance. Glycogen storage due to increased G-6-P levels.
VIII	mild hypoglycemia	Kinase inactive in phosphorylating liver glycogen phosphorylase, thereby disabling glycogen degradation.

Table 4.3. Glycogen Storage Diseases: Clinical Features and Mechanisms

Pyruvate dehydrogenase is a complex of several enzymes and cofactors. Depending upon the source, the multienzyme complex will consist of 20 - 30 **pyruvate dehydrogenase subunits**, 60 **dihydrolipoyl transacetylase subunits** and 5-6 **dihydrolipoyl dehydrogenase subunits**. This structural integration of three enzymes into a large complex facilitates the coordinated catalysis of this complex, multistep reaction. A very similar complex catalyzes the <u>α-ketoglutarate dehydrogenase</u> reaction which occurs within the Citric Acid Cycle (see below) and a step in the metabolism of leucine. The actions and functions of the complex are diagrammed in figure 4.2.

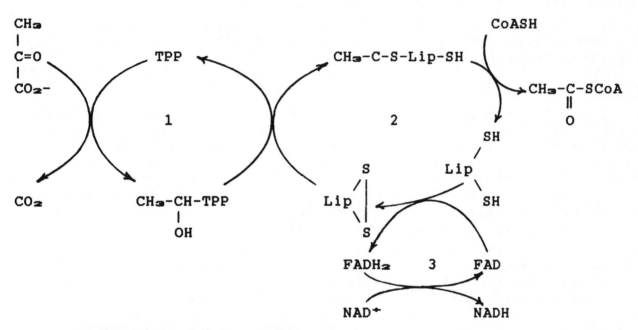

1. Pyruvate dehydrogenase
2. Dihydrolipoyl transacetylase
3. Dihydrolipoyl dehydrogenase

<u>Figure 4.2</u>. The Pyruvate Dehydrogenase Complex

The cofactors involved in the pyruvate dehydrogenase reaction are:

TPP: thiamine pyrophosphate.

Lip: lipoamide; lipoic acid attached to a specific lysine side chain of <u>dihydrolipoyl transacetylase</u>. Active groups are two sulfhydryl groups which are either reduced (SH) or oxidized (S-S).

CoA: Coenzyme A, which exists either as the free sulfhydryl (CoASH) or the acetylated (Acetyl CoA) form. The latter is the metabolically active form of acetate.

FAD: Flavin Adenine Dinucleotide. $FAD/FADH_2$ is a hydrogen receptor/donor (like NAD^+) which generally operates at lower potentials than the $NAD^+/NADH$ system. In this reaction FAD is a tightly bound prosthetic group on the **dihydrolipoyl dehydrogenase** enzyme.

There are four steps in this reaction:

1. Pyruvate is bound to a reactive C-atom in TPP through its α-keto group and decarboxylated, leaving hydroxyethyl TPP. This reaction is catalyzed by the pyruvate dehydrogenase subunit.

2. The hydroxyethyl group is oxidized to an acetyl group and transferred to Lipoamide. The oxidant is the disulfide group of Lipoamide, which is converted to sulfhydryl groups in the process. This reaction is catalyzed by the dihydrolipoyl transacetylase subunit.

3. The acetyl group is transferred from lipoamide to CoASH, also by the dihydrolipoyl transacetylase subunit.

4. The lipoamide sulfhydryl groups are regenerated. NADH is the proton donor but acts through an FAD prosthetic group on the dihydrolipoyl dehydrogenase subunit.

B. The Citric Acid Cycle

The cycle is shown in Figure 4.3. A four-carbon compound (oxaloacetate) condenses with a two-carbon acetyl moiety to form the six-carbon compound, citrate, which is then isomerized and oxidatively decarboxylated. The resulting five-carbon compound (α-ketoglutarate) is oxidatively decarboxylated, yielding a four-carbon moiety, succinyl CoA. The energy in the CoA derivative is captured in the form of GTP when succinate is released. Succinate is then converted into oxaloacetate in three steps involving two oxidations.

For each turn of the cycle, two carbons enter the cycle as acetate while two leave as CO_2. However, though the stoichiometry suggests that the same carbon atoms are involved, this is not so. The carbon atoms which are eliminated as CO_2 are the two carboxylate carbons present on oxaloacetate prior to condensation with acetyl CoA. In addition to the carbon atoms leaving the cycle, four pairs of hydrogen atoms leave in the oxidation steps, three pairs in the form of NADH and one in the form of $FADH_2$. The reduced nucleotides are passed directly to the electron transport system where oxidative phosphorylation converts them to ATP. Because of the necessity of metabolizing the reduced nucleotides, *the citric acid cycle is only active under aerobic conditions*.

The cycle is also a source of biosynthetic intermediates as well as a means of metabolizing other compounds. α-ketoglutarate, pyruvate and oxaloacetate connect directly with amino acid metabolism, while most of the carbon atoms in porphyrins are contributed by succinyl CoA synthesized by the cycle. Fatty acids and a number of amino acids yield acetyl CoA as end products of their metabolism. However, citric acid cycle intermediates which are drawn off for biosynthesis must be replaced if the cycle is to continue functioning. Mammals lack the enzymatic machinery to convert acetyl CoA into oxaloacetate or any of the other intermediates in the cycle. Mammals replenish oxaloacetate from pyruvate using pyruvate carboxylase, an enzyme which was discussed earlier in gluconeogenesis.

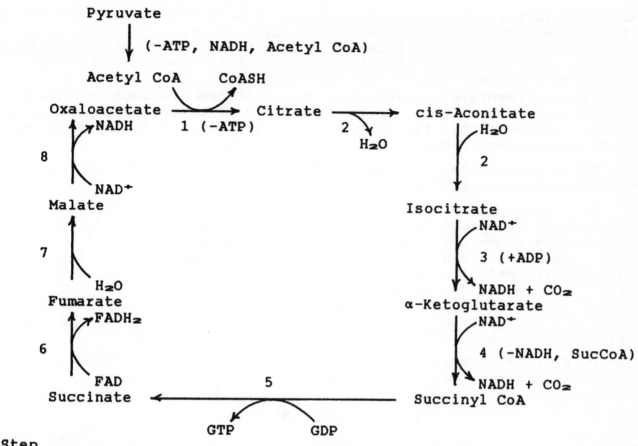

Step Nbr	Enzyme	Cofactor	Type*
1	Citrate synthetase	CoA	C
2	Aconitase	Fe^{++}	D/H
3	Isocitrate dehydrogenase	NAD^+, CoA	DC/O
4	α-ketoglutarate dehydrogenase complex	TPP, Lipoamide, FAD	DC/O
5	Succinyl CoA synthetase	CoA	P
6	Succinate dehydrogenase	FAD (enzyme bound)	O
7	Fumarase	None	H
8	Malate dehydrogenase	NAD^+	O

<u>Figure 4.3</u>. The Citric Acid Cycle. *C = condensation, D = dehydration, DC = decarboxylation, H = hydration, O = oxidation, P = phosphorylation.

C. <u>Regulation of the Citric Acid Cycle</u>

Entry of carbohydrate into the cycle is tied closely to the needs for energy metabolism and levels of key metabolites. The cycle is inactive when NADH, ATP and acetyl CoA are plentiful, but active when they are scarce and when pyruvate is plentiful.

1. Supply of substrates. The availability of acetyl CoA (whether from fatty acids or pyruvate, i.e. carbohydrate), oxaloacetate, and both NAD^+ and FAD are major determinants of the activity of the cycle.

Note that in the absence of pyruvate, which is provided primarily by carbohydrate metabolism, the unavailability of oxaloacetate resulting from the draining off of intermediates for biosynthesis will virtually shut down the cycle and simultaneously inhibit metabolism of fatty acids. This is the origin of the aphorism that "fats burn in the flame of carbohydrate". The need for oxidized nucleotides closely couples cycle activity to generation of ATP in oxidative phosphorylation and the availability of ADP, P_i, and water. This is the origin of the "respiratory control" of the cycle.

2. Effector-mediated control. Four steps are involved in allosteric control of the cycle:

a. _Pyruvate dehydrogenase_ is inhibited by its products, acetyl CoA (inhibits transacetylase subunit) and NADH (inhibits dihydrolipoyl dehydrogenase subunit), but the inhibition is reversed by high levels of the reactants, CoASH and NAD^+. The pyruvate dehydrogenase subunit is also allosterically inhibited by GTP but activated by AMP. Finally, the pyruvate dehydrogenase subunit is also inactivated by covalent modification, the phosphorylation of a specific serine residue by ATP. This phosphorylation is enhanced by high ratios of ATP/ADP, acetyl CoA/CoASH and NADH/NAD^+, but inhibited by pyruvate.

b. _Citrate synthetase_ is inhibited allosterically by ATP, which reduces its affinity for acetyl CoA.

c. _Isocitrate dehydrogenase_ is allosterically stimulated by ADP but inhibited by NADH, which displaces NAD^+.

d. _α-Ketoglutarate dehydrogenase_ is regulated by mechanisms similar to those which regulate pyruvate dehydrogenase. Succinyl CoA, NADH and a high energy charge inhibit the complex.

VII. GLYCOCONJUGATES OR COMPLEX CARBOHYDRATES

Carbohydrates frequently exist as complex structures of the following types.

A. _Glycoproteins_: Proteins which have oligosaccharides covalently attached to the protein chain. Virtually all membrane and plasma proteins are glycoproteins, while cytoplasmic proteins frequently do not contain carbohydrate. The oligosaccharides serve as recognition markers and receptors. The carbohydrate chains of plasma proteins usually contain a terminal sialic acid and its removal usually causes the protein to be rapidly cleared by the liver. _Sugars which are found in glycoproteins include; glucose, galactose, mannose, N-acetyl galactosamine, N-acetyl glucosamine, fucose, arabinose, xylose and N-acetyl neuraminic acid, or sialic acid._

B. _Glycolipids_: Lipids containing a covalently attached oligosaccharide and which are usually found in membranes with the carbohydrate moiety pointing outward. Many cell-surface antigens or recognition markers are glycolipids. The oligosaccharides are similar in structure to those found in glycoproteins.

C. <u>Proteoglycans</u>:   Complex   molecules composed of a core protein with covalently attached glycosaminoglycans (see below). The main difference between proteoglycans and glycoproteins is the nature of the attached carbohydrate.   Glycosaminoglycans are higher molecular weight, anionic polymers, while the oligosaccharides of glycoproteins are composed predominantly of neutral sugars and are usually smaller. Proteoglycans, particularly those containing chondroitin sulfates, hyaluronate and dermatan sulfate, are major structural elements of connective tissue. Basement membrane and the cell surface are rich in heparin sulfate-containing proteoglycans where they serve important, if poorly understood, functions.   Heparin proteoglycan is found in mast cells and basophils, where it serves to store histamine.

D. <u>Glycosaminoglycans</u>:   Linear polysaccharides generally composed of alternating hexosamine and either uronic acid or galactose (in keratan sulfate only).   Because of this alternating structure, the basic structural unit is a disaccharide. All but hyaluronate contain sulfated sugars. Glycosaminoglycans are usually found as constituents of proteoglycans, though heparin is active as an anticoagulant as free chains.   *The sugars found in glycosaminoglycans are: N-acetyl glucosamine (glcNAc) and -galactosamine (gal NAc); glucuronic (GlcU) and iduronic acids (IdU), and variously sulfated derivatives such as N-acetyl galactosamine 4-sulfate or 6-sulfate, and the like.*

IX. SOME IMPORTANT STRUCTURES

Most simple sugars can be related to the structure of D-glucose. Its structure can be expressed by three major conventions: Fisher formulas, Haworth formulas, and stereochemical projections. A mnemonic device can be derived for each to help remember the proper stereochemical relationships.   For the Fisher formulas, glucose is "RLRR", "R" being "right" and "L" being "left". Modern texts use the Haworth convention to show the ring forms. With the Haworth formulas, "U" (up) and "D" (down) must be remembered with "DUDU" giving the configurations for carbons 2,3,4 and 6 of glucose. The rule for the anomeric carbon (i.e. the one derived from the carbonyl function) of D-sugars is "ADD" (alpha, <u>D</u>, <u>D</u>own), with the reverse being true for the L-sugars.   For the stereochemical projections, all of the substituents of ß-D-glucose are equatorial (e; parallel to the equator of the ring and perpendicular to the plane of the paper), as opposed to axial (a; parallel to the axis of the ring and the plane of the paper).

<u>**Fisher formula**</u>   <u>**Haworth formula**</u>   <u>**Stereochemical formula**</u>

D-Glucose   α-D-glucose   OH   ß-D-glucose

Most of the biologically significant monosaccharides are <u>epimers</u> of glucose, that is they differ by the configuration about a single carbon atom. Thus, galactose is the 4-epimer of glucose while mannose is the 2-epimer. Each of these can be drawn or remembered by inverting (change R to L and D to U, or vice versa) the configuration about the appropriate carbon atom.

The most significant ketohexose is fructose, which has the same configuration as glucose but has a keto group on C-2. The Haworth formula of the α anomer is:

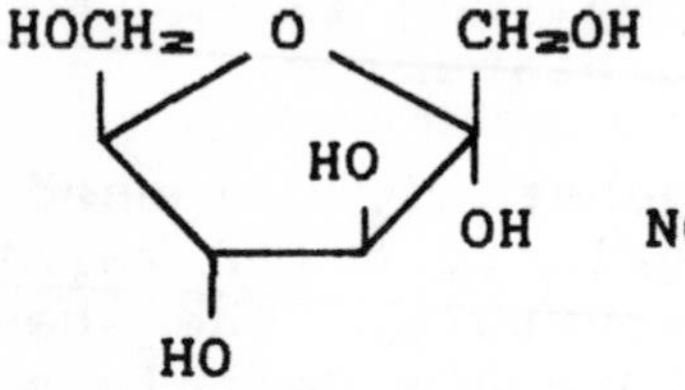
HOCH₂
O
CH₂OH
HO
OH
HO

NOTE: "ADD" holds for the anomeric C-atom

Disaccharides consist of two sugars bound together. At least one sugar will be bound through the anomeric carbon atom. This is a <u>glycosidic bond</u>. If only one sugar is bound through its anomeric carbon, then the sugar is "reducing", since the anomeric carbon of the other is free and, hence, capable of forming the free aldehyde group. The latter reduces alkaline copper reagents. If the anomeric carbon atoms are bound to each other, then the sugar is "non-reducing", since neither can yield a free aldehyde group. Lactose and sucrose shown below are reducing and non-reducing disaccharides respectively.

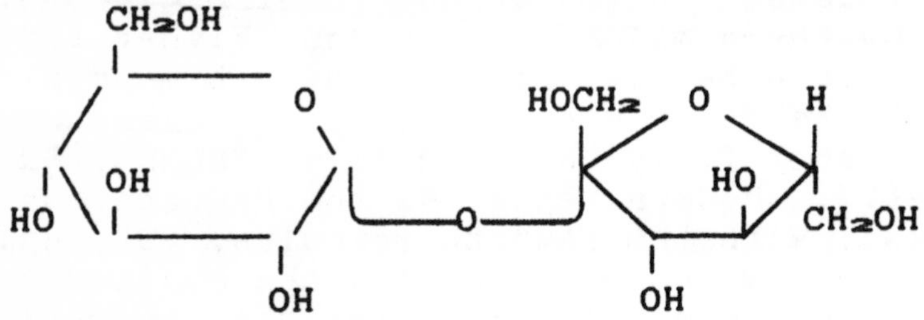
CH₂OH
O
HOCH₂
O
H
OH
HO
HO
O
CH₂OH
OH
OH

Sucrose, or α-D-glucose-1,1-α-D-fructose

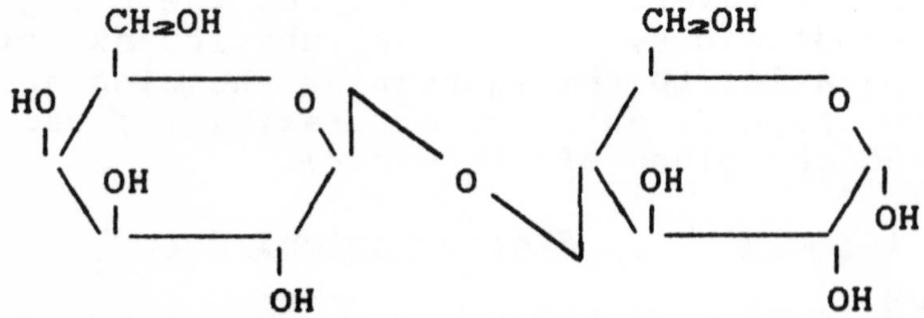
CH₂OH
CH₂OH
HO
O
O
OH
O
OH
OH
OH
OH

α-Lactose, or ß-D-galactose-1,4-α-D-glucose

Note that lactose can have both α and ß anomers, depending upon the configuration of the glucose, while sucrose cannot have anomers, lacking an anomeric carbon atom.

IX. REVIEW QUESTIONS ON CARBOHYDRATE METABOLISM

DIRECTIONS: Each of the questions or incomplete statements below is followed by five suggested answers or completions. Select the one that is BEST in each case and fill in the corresponding space on the answer sheet.

1. Glycolysis in muscle is reduced when fatty acid oxidation is increased because:

A. oxidation of fatty acids increases the formation of citrate which blocks phosphofructokinase.
B. utilization of fatty acid increases the level of NADPH which decreases the activity of hexokinase.
C. fatty acids compete with glucose for insulin-mediated transport across the cell membrane.
D. fatty acids inhibit the formation of pyruvate from lactate.
E. fatty acids compete with glucose for NAD^+, which blocks the formation of pyruvate.

2. In mammals each of the following is a function of the tricarboxylic acid cycle EXCEPT:

A. net synthesis of oxaloacetate from acetyl CoA
B. formation of α-ketoglutarate for amino acid biosynthesis.
C. generation of NADH and $FADH_2$.
D. metabolism of acetate to carbon dioxide and water.
E. oxidation of acetyl CoA produced primarily from glycolysis and oxidation of fatty acids.

3. When a normal person consumes a low-carbohydrate diet, the liver produces much greater amounts of ketone bodies because:

A. sufficient α-glycerol phosphate is not present
B. the pentose phosphate pathway fails to produce sufficient NADPH
C. blood insulin levels are lower
D. there is excess NADH resulting from fatty acid oxidation
E. the TCA cycle fails to keep pace with the production of acetyl CoA by oxidation of fatty acids

4. An individual who cannot synthesize functional liver fructose 1,6-bisphosphatase would be primarily affected by:

A. a failure to resynthesize glucose from lactate produced during exercise
B. an inability to metabolize fructose
C. a lowered yield of ATP production per mole of glucose metabolized
D. a failure to split fructose diphosphates into triose phosphates
E. none of the above

5. During starvation, as gluconeogenesis increases to maintain the levels of blood glucose, which one of the following will be enhanced?

A. liver pyruvate kinase activity
B. the secretion of insulin by the pancreas
C. muscle glucose-6-phosphatase activity
D. the metabolism of acetyl CoA to pyruvate
E. the metabolism of glutamate to glucose-6-phosphate

6. When one mole of glycerol is metabolized completely to CO_2 and water by the glycolytic pathway and the TCA cycle, what is the net maximum yield of ATP?

A. 10-14
B. 15-19
C. 20-24
D. 25-30
E. 31-36

7. When glucose provides the raw material for the synthesis of fatty acids, how many carbon atoms end up in fatty acid?

A. 2
B. 3
C. 4
D. 5
E. 6

8. In humans, which of the following enzyme-catalyzed reactions does NOT produce CO_2?

A. isocitrate dehydrogenase
B. pyruvate dehydrogenase
C. α-ketoglutarate dehydrogenase
D. 6-phosphogluconate dehydrogenase
E. succinate dehydrogenase

9. Biotin is required as a coenzyme in which one of the following reactions?

A. α-ketoglutarate + NAD$^+$ + CoA $\longrightarrow$
      succinyl CoA + CO_2 + NADH
B. pyruvate + CO_2 + ATP $\longrightarrow$
      oxaloacetate + ADP + P_i
C. pyruvate + NAD$^+$ + CoA $\longrightarrow$
      acetyl CoA + CO_2 + NADH
D. 6-phosphogluconate $\longrightarrow$
      ribulose-5-phosphate + CO_2
E. α-ketoglutarate + CO_2 + NADH $\longrightarrow$
      isocitrate + NAD$^+$

10. In addition to an enzyme complex, the conversion of pyruvate to acetyl CoA requires:

A. CoA, thiamine pyrophosphate and NAD$^+$
B. CoA, lipoic acid, thiamine pyrophosphate and FAD
C. CoA, lipoic acid, thiamine pyrophosphate and NAD$^+$
D. CoA, lipoic acid, biotin and ATP
E. CoA, ATP, NAD$^+$ and riboflavin

11. In hereditary fructose intolerance, the primary biochemical defect is:

A. a deficiency in activity of an aldolase isozyme
B. increased allosteric sensitivity of phosphofructokinase to AMP
C. inhibition of glycogen synthetase
D. an inability to absorb fructose
E. a deficiency in the activity of fructokinase

12. Hyaluronic acid is a:

A. glycoprotein
B. high molecular weight, positively charged polysaccharide
C. polymer which contains sulfate
D. repeating disaccharide of glucuronic acid and N-acetylglucosamine
E. all of the above

13. Sialic acid is:

A. found only in mammalian tissues
B. the major carbohydrate found in heparin
C. a normal constituent of glycoproteins
D. an ε-carboxy amino acid
E. a cofactor for neuraminidase

14. Which one of the following is required for the conversion of succinate to fumarate?

A. ATP
B. NAD$^+$
C. NADP$^+$
D. biotin
E. FAD

15. A patient presenting with a suspected metabolic disorder shows (1) abnormally high amounts of glycogen with normal structure in liver, and (2) no increase in blood glucose levels following oral administration of fructose. From these two findings, which one of the following enzymes is likely to be deficient?

A. phosphoglucomutase
B. UDP-glycogen transglucosylase
C. fructokinase
D. glucose-6-phosphatase
E. glucokinase

16. The saccharide containing the largest number of glucose units per mole is:

A. maltose
B. trehalose
C. sucrose
D. amylase
E. glycogen

17. Glucose, labeled with ^{14}C in different carbon atoms is added to a tissue that is rich in the enzymes of the hexose monophosphate shunt. Which one will give the most rapid initial evolution of $^{14}CO_2$?

A. glucose-1-^{14}C
B. glucose-2-^{14}C
C. glucose-3,4-^{14}C
D. glucose-5-^{14}C
E. glucose-6-^{14}C

18. The tissue with the lowest activity for the oxidation of glucose-6-phosphate by the phosphogluconate pathway is:

A. liver
B. lactating mammary gland
C. striated muscle
D. adrenal cortex
E. adipose tissue

19. The absence of which one of the following reactions is responsible for the inability of man to use fatty acids in the de novo net synthesis of glucose?

A. oxaloacetate $\longrightarrow$ pyruvate
B. oxaloacetate + acetyl CoA $\longrightarrow$ citrate
C. acetyl CoA $\longrightarrow$ pyruvate
D. pyruvate $\longrightarrow$ phosphoenolpyruvate
E. phosphoenolpyruvate $\longrightarrow$ oxaloacetate

20. Which one of the following substrates can NOT contribute to net gluconeogenesis in mammalian liver?

A. alanine
B. stearate
C. α-ketoglutarate
D. glutamate
E. pyruvate

21. Which one of the following enzymes contains lipoic acid bound in an amide linkage?

A. lactate dehydrogenase
B. phosphofructokinase
C. glycogen synthetase
D. ferrochelatase
E. pyruvate dehydrogenase

22. Transketolase requires which one of the following coenzymes?

A. pyridoxal phosphate
B. lipoamide
C. thiamine pyrophosphate
D. cobalamin
E. tetrahydrofolic acid

23. Epinephrine:

A. is a cofactor for protein kinase
B. regulates glycogen metabolism by inhibiting phosphorylase and activating glycogen synthetase
C. stimulates membrane-bound adenylate cyclase to produce cAMP
D. increases glucose levels and decreases free fatty acid levels in blood
E. acts predominantly on liver cells

24. Conversion of phosphorylase b to phosphorylase a involves a reaction with which amino acid in the enzyme's polypeptide backbone?

A. aspartic acid
B. serine
C. glycine
D. cysteine
E. arginine

25. 3-phosphoglyceraldehyde dehydrogenase produces which one of the following as a product?

A. inosine triphosphate
B. 1,3-diphosphoglycerate
C. cytidine triphosphate
D. phosphoenolpyruvate
E. phosphocreatine

26. Which one of the following participates directly in both oxidation-reduction reactions in glycolysis?

A. ADP
B. creatine phosphate
C. glyceraldehyde-3-phosphate
D. NAD+
E. FAD

27. Unusually high concentrations of
liver glycogen could be produced by:

A. adrenocortical insufficiency
B. a lack of insulin
C. a lack of glycogen synthetase
D. alkalosis
E. a lack of phosphorylase kinase

28. Conversion of glycogen synthetase a
(active) to glycogen synthetase b
(inactive) involves:

A. an increased synthesis of phos-
   phorylase
B. decreased cAMP
C. phosphorylation of the enzyme
D. the concentration of glucose-6-phos-
   phate
E. dephosphorylation of the enzyme

29. In the metabolism of glycerol to
glycogen, the first intermediate of glycolysis encountered is:

A. glycerol 3-phosphate
B. dihydroxyacetone phosphate
C. 3-phosphoglycerate
D. ribulose-5-phosphate
E. 1,3-bisphosphoglycerate

30. The enzyme at which glycolysis is
regulated under anaerobic conditions is:

A. pyruvate kinase
B. phosphofructokinase
C. phosphoglyceromutase
D. triose phosphate isomerase
E. fructose-1,6-diphosphate

31. Collagen contains a carbohydrate
moiety linked to:

A. threonine
B. hydroxyproline
C. hydroxylysine
D. asparagine
E. serine

32. A mouse was fed glucose labeled in
the C-1 position with ^{14}C. The following
amino acids were isolated from hy-
drolysates of tissue proteins. Which
one was not labeled?

A. glutamate
B. asparagine
C. cysteine
D. leucine
E. proline

33. Acetyl CoA does not pass from the
mitochondria to the cytoplasm. This occurs after acetyl CoA is converted to:

A. citrate
B. acetate
C. pyruvate
D. fatty acid
E. oxaloacetate

34. Uniformly labeled ^{14}C-oxaloacetate
is condensed with unlabeled acetyl CoA.
After a single turn around the tricarboxylic acid cycle back to oxaloacetate,
what fraction of the original radioac-
tivity will be found in the oxaloacetate?

A. all
B. 3/4
C. 1/2
D. 1/4
E. 1/3

DIRECTIONS: For each of the questions or incomplete statements below, ONE or MORE of the answers or completions is correct. On the answer sheet fill in space

A if only 1, 2, and 3 are correct
B if only 1 and 3 are correct
C if only 2 and 4 are correct
D if only 4 is correct
E if all are correct

FILL IN ONLY ONE SPACE ON YOUR ANSWER SHEET FOR EACH QUESTION

Directions Summarized				
(A) 1,2,3 only	(B) 1,3 only	(C) 2,4 only	(D) 4 only	(E) All are correct

35. Glycosaminoglycans are characterized by:

1. a linear chain
2. multiple anionic sites
3. ester sulfate
4. the presences of N-acetyl neuraminic acid

36. Activated sugar residues utilized for the biosynthesis of complex carbohydrates include:

1. GDP-mannose
2. dolichol phosphorylglucose
3. UDP-glucuronic acid
4. CDP-N-acetylneuraminic acid

37. Products of reactions in which CO_2 fixation occurs by means of enzyme-bound biotin include:

1. phosphoenolpyruvate
2. 6-phosphogluconate
3. succinyl CoA
4. oxaloacetate

38. During the period of no food intake between the evening meal and breakfast the next day, the brain:

1. derives most of its energy from the oxidation of amino acids.
2. metabolizes its own stores of glycogen.
3. metabolizes acetoacetate
4. uses blood glucose derived from the breakdown of hepatic glycogen.

39. Enzymatic hydrolysis of starch (amylose) by amylase:

1. occurs in the stomach and small intestine
2. yields products that contain 1,4- and 1,6-linked glucose chains
3. involves an enzyme released by the salivary glands and the pancreas
4. Yields products that are absorbed in the large intestine.

40. Possible causes of hypoglycemia include:

1. an anterior pituitary tumor producing increased amounts of growth hormone.
2. a deficiency of glucagon
3. a hypothalamic tumor producing increased amounts of ACTH
4. a pancreatic islet tumor producing increased amounts of insulin

41. Galactose:

1. is a component of some glycoproteins
2. results from the hydrolysis of lactose
3. is converted by a cyclic metabolic pathway to glucose-1-phosphate
4. cannot be metabolized by diabetics

42. The activity of glycogen phosphorylase in muscle is regulated by:

1. glucagon
2. phosphorylase kinase a
3. glucose-1-phosphate
4. epinephrine

FILL IN ONLY ONE SPACE ON YOUR ANSWER SHEET FOR EACH QUESTION

Directions Summarized				
(A)	(B)	(C)	(D)	(E)
1,2,3	1,3	2,4	4	All are
only	only	only	only	correct

43. A normal person has received a large dose of insulin. Which of the following events will occur?

1. increased epinephrine secretion
2. a decrease in blood glucose
3. increased glycogen synthesis
4. increased uptake of amino acids by muscle

44. Glucagon

1. has actions similar to those of insulin
2. is secreted by the alpha cells of the pancreatic islets
3. decreases cyclic AMP concentrations in many tissues
4. targets liver primarily

45 Glycolysis in the red blood cell produces:

1. lactic acid
2. NADH
3. ATP
4. CO_2

46. The increase of glycogenolysis in muscle produced by epinephrine may be attributed to:

1. decreased Ca^{++}
2. activation of aldolase
3. reduction in total NAD^+ plus NADH
4. conversion of phosphorylase b to phosphorylase a

47. An increase in the ATP/ADP ratio in liver cells would lead to a decrease in the activity of:

1. phosphofructokinase
2. fructose 1,6-bisphosphatase
3. NAD^+-linked isocitrate dehydrogenase
4. citrate lyase

48. Hydrolytic anemia associated with a deficiency of erythrocyte glucose-6-phosphate dehydrogenase is marked by:

1. increased accumulation of lipid hydroperoxides
2. exacerbation by drugs which are usually harmless
3. increased ratios of oxidized to reduced glutathione
4. a high ratio of NAD^+ to NADPH

49. In humans, gluconeogenesis:

1. is important in the conversion of fatty acid to glucose
2. involves the formation of phosphoenol pyruvate from oxaloacetate
3. helps maintain blood glucose concentration after a starch-rich meal
4. results in the conversion of protein to blood glucose

50. Major mechanisms by which the glycolytic pathway is regulated include:

1. inhibition of pyruvate dehydrogenase by ATP
2. inhibition of phosphofructokinase by citrate
3. stimulation of phosphofructokinase by AMP
4. stimulation of pyruvate kinase by cyclic AMP-dependent protein kinase

51. The oxidative branch of the pentose pathway:

1. uses glucose-6-phosphate as the substrate in the first reaction
2. provides a product which helps maintain erythrocyte membrane stability
3. provides the reductant needed for the synthesis of fatty acids
4. is essential for the synthesis of pentoses

FILL IN ONLY ONE SPACE ON YOUR ANSWER SHEET FOR EACH QUESTION

Directions Summarized				
(A) 1,2,3 only	(B) 1,3 only	(C) 2,4 only	(D) 4 only	(E) All are correct

52. Glycoproteins:

1. contain sugars linked to asparagine, serine, threonine or hydroxylysine
2. require removal of sialic acid to prevent clearance from the circulation
3. are first synthesized as polypeptides followed by the addition of carbohydrate
4. are nearly always fibrous proteins

53. Glucose-1-phosphate is formed in the liver by the action of:

1. phosphoglucomutase on glucose-6-phosphate
2. a transferase on UDP-glucose and galactose-1-phosphate
3. cAMP-dependent protein kinase on glycogen
4. an epimerase on fructose-1-phosphate

54. Anoxia increases

1. the activity of hexokinase
2. the concentration of ATP
3. the rate of the citric acid cycle
4. the rate of glycolysis

55. Citrate:

1. stimulates synthesis of fatty acids
2. activates acetyl CoA carboxylase
3. regulates glycolysis by inhibiting phosphofructokinase
4. acts to transport acetyl CoA to the cytosol from the mitosol

56. 2,3,-Bisphosphoglycerate:

1. is an intermediate in the phosphoglyceromutase reaction
2. is an intermediate in the glycolytic pathway
3. reduces the affinity of hemoglobin for O_2
4. readily acts as a phosphoryl donor

57. Epinephrine (in muscle) and glucagon (in liver):

1. activate adenyl cyclase
2. inactivate phosphorylase and activate glycogen synthetase
3. stimulate glycogenolysis
4. stimulate glycogen synthesis

58. Enzymes present in liver and kidney which are necessary for gluconeogenesis include:

1. glucose-6-phosphate dehydrogenase
2. glucose-6-phosphatase
3. phosphofructokinase
4. pyruvate carboxylase

59. Acetyl CoA oxidation leads to

1. net consumption of 1 mole of oxaloacetate
2. net production of 1 mole of citrate
3. net production of 6 ATP
4. net production of 2 CO_2

DIRECTIONS: Each group of items below consists of lettered headings followed by a set of numbered words or phrases. For each numbered word or phrase, select the ONE heading that is most closely associated with it and fill in the corresponding space on the answer sheet. Each heading may be used once, more than once, or not at all.

Questions 60-62:

A. ATP is a substrate
B. ATP is an inhibitor
C. AMP is an inhibitor
D. ATP is both a substrate and an inhibitor

60. Phosphofructokinase

61. glucokinase

62. fructose-1,6-bisphosphatase

Questions 63-64:

A. FAD
B. pyridoxal phosphate
C. NAD$^+$
D. thiamine pyrophosphate
E. all of the above

63. Forms an acyl derivative during the enzymatic breakdown of pyruvate

64. prosthetic group of the pyruvate dehydrogenase complex

Questions 65-68:

A. lipoyl lysine (lipoic acid).
B. NAD$^+$
C. NADP$^+$
D. biotin
E. any two of the above

65. Involved in the conversion of glyceraldehyde-3-P to 1,3-diphosphoglycerate

66. Involved in fatty acid biosynthesis

67. Involved in the conversion of glucose-6-P to 6-phosphogluconate

68. Involved in the oxidation of pyruvate

Questions 69-72:

A. glycerol $\longrightarrow$ glucose
B. pyruvate $\longrightarrow$ acetyl CoA
C. glucose $\longrightarrow$ fatty acid
D. glucose $\longrightarrow$ lactate
E. glucose $\longrightarrow$ fructose

In mammalian metabolism the pathway that:

69. Requires the operation of the hexose monophosphate shunt

70. Cannot be reversed to achieve net synthesis of glucose

71. Yields a net synthesis of two ATP

72. Allows triglycerides to contribute to glucose synthesis

Questions 73-76:

A. glucose-6-P dehydrogenase deficiency
B. glucose-6-phosphatase deficiency
C. galactokinase deficiency
D. UDP-galactose epimerase deficiency
E. hexokinase deficiency

73. Increased concentration of galactose-1-P

74. Increased lipid peroxides in erythrocytes

75. Increased concentration of liver glycogen

76. Essentially benign excretion of a glucose epimer

DIRECTIONS: Each group of items below consists of lettered headings followed by a set of numbered words or phrases. For each numbered word or phrase, select the ONE heading that is most closely associated with it and fill in the corresponding space on the answer sheet. Each heading may be used once, more than once, or not at all.

Questions 77-79:

    A. glucose-6-phosphate
    B. UDP-galactose
    C. lactate
    D. acetyl CoA
    E. 1,3-bisphosphoglycerate

77. Glycogen synthetase

78. Pyruvate carboxylase

79. Cannot be converted to glucose

Questions 80-82:

    A. galactose
    B. glucose
    C. fucose
    D. N-acetylgalactosamine
    E. glucosamine

80. Dextran

81. Lactose

82. Hyaluronic acid

DIRECTIONS: Each set of lettered headings below is followed by a list of numbered words or phrases. For each numbered word or phrase select:

    A if the item is associated with (A) only
    B if the item is associated with (B) only
    C if the item is associated with both (A) and (B)
    D if the item is associated with neither (A) nor (B)

Questions 83-87:

    A. synthesis of glycogen from glucose
    B. degradation of glycogen to glucose
    C. both
    D. neither

83. Requires UDP-galactose

84. Requires inorganic phosphate

85. Stimulated by an increase in insulin

86. Glucose-1-phosphate is an intermediate

87. Inhibited by an increase in cAMP

Questions 88-90:

    A. requires nucleoside triphosphate
    B. generates nucleoside triphosphate
    C. both
    D. neither

88. Glucose $\longrightarrow$ glycogen

89. Glucose $\longrightarrow$ pyruvate

90. Glyceraldehyde-3-phosphate $\longrightarrow$ 1,3-bisphosphoglycerate

X. ANSWERS TO QUESTIONS ON CARBOHYDRATES

1. A	26. D	51. A	76. C
2. A	27. E	52. B	77. A
3. E	28. C	53. A	78. D
4. A	29. B	54. D	79. D
5. E	30. B	55. E	80. B
6. C	31. C	56. A	81. A
7. C	32. D	57. B	82. D
8. E	33. A	58. C	83. D
9. B	34. C	59. D	84. B
10. C	35. A	60. D	85. A
11. A	36. B	61. A	86. C
12. D	37. D	62. C	87. A
13. C	38. D	63. D	88. A
14. E	39. A	64. E	89. C
15. D	40. C	65. B	90. A
16. E	41. A	66. E	
17. A	42. C	67. C	
18. C	43. E	68. E	
19. C	44. C	69. C	
20. B	45. A	70. B	
21. E	46. D	71. D	
22. C	47. A	72. A	
23. C	48. E	73. D	
24. B	49. C	74. A	
25. B	50. A	75. B	

5. AMINO ACID METABOLISM

A. M. Chandler

I. FUNCTIONS OF AMINO ACIDS IN MAN

Amino acids are ingested in large amounts as structural components of dietary proteins. Unlike carbohydrate and fat, there are no large reserve stores of protein in the body. Thus, a continuous intake is required if tissue breakdown is to be avoided.

Amino acids are:

1. Precursors for the synthesis of proteins.
2. A source of energy under certain conditions.
3. Involved in the detoxification of drugs, chemicals and metabolic by-products.
4. Involved as direct neurotransmitters or as precursors to neurotransmitters.
5. Precursors to several peptide hormones and thyroid hormone.
6. Precursors to histamine, NAD and miscellaneous compounds of biological importance.

II. ESSENTIAL AND NON-ESSENTIAL AMINO ACIDS

All twenty amino acids are essential for life. A lack of a sufficient amount of any one of them leads to severe metabolic disruption and ultimate death.

Most microorganisms and plants are able to synthesize all 20 from glucose or CO_2 and NH_3. Mammals, however, including man, have during the process of evolution lost the ability to synthesize the carbon skeletons for several of the amino acids. Therefore, it is _essential_ that these particular amino acids be obtained through the diet. Those amino acids that are not synthesized at a sufficient rate to meet demand are termed the **essential amino acids** and for man number ten. A useful mnemonic to assist in remembering the essential amino acids is PVT TIM HALL:

P- phenylalanine	T- tryptophan	H- histidine*
V- valine	I- isoleucine	A- arginine *
T- threonine	M- methionine	L- lysine
		L- leucine

Note that histidine and arginine are marked with asterisks. These amino acids are undoubtedly required for the infant and growing child, but it is less clear that they are essential for the normal, healthy adult.

III. NITROGEN BALANCE

The greatest portion of N intake is in the form of amino acids in the protein of the diet. After digestion, absorption and metabolic processing, the excess N derived from the NH_2 not required for growth or maintenance is excreted in the urine in the form of urea, NH_3 and other nitrogenous compounds. The normal, healthy adult is in "**nitrogen balance**" or "**equilibrium**". That is, the amount of N ingested in the diet over a given period of time equals that excreted in the urine and feces as excretory products.

Positive and Negative Nitrogen Balance: During pregnancy, infancy, childhood and when in the recovery phase from a severe illness or surgery, the amount of N taken in and retained exceeds that excreted. The organism is said to be in a state of **positive nitrogen balance**. On the other hand, during starvation, <u>immediately</u> following severe trauma, surgery or other acute stress such as infections, N excretion exceeds intake and retention and the organism is in a state of **negative nitrogen balance**. A gradual, prolonged negative N balance is associated with **senescence**. Nitrogen balance is humorally controlled. Positive N balance is associated with growth hormone, insulin and with testosterone and other anabolic steroids. Negative N balance is associated with glucocorticoid action in mobilizing amino acids from muscle tissue.

IV. PROTEIN QUALITY

The dietary source of protein is also important for maintaining N balance. Not all proteins have the same biological value (BV). Proteins derived from animal sources have a high BV because they contain all the essential amino acids in the proper proportions. Plant proteins, on the other hand, usually are in lower tissue concentration and are harder to digest. In general, plant proteins are deficient in one or more essential amino acids, primarily lysine, tryptophan or methionine. In some third-world countries animal proteins are almost non-existent and protein intake may be limited to only one or two plant sources. The lack of a single essential amino acid leads to severe growth retardation in children and in adults to negative N balance. Growing children are particularly vulnerable and this is evidenced by the prevalence of **kwashiorkor** (See Chapter 12). Strict vegetarians can do well if they plan a diet containing a mixture of vegetable proteins, each one compensating for a defect in the other.

V. PROTEIN DIGESTION

A. Gastric Digestion

The first phase of protein digestion takes place in the stomach. **Gastrin**, a polypeptide hormone, is secreted into the blood by the antral gastric mucosa upon stimulation by foods. Ethanol is a particularly strong stimulator of gastrin release. Gastrin stimulates **chief**

cells of the gastric mucosa to secrete the inactive proenzyme, **pepsinogen**, the **parietal cells** to secrete **HCL** and the **epithelial cells** to secrete **mucoproteins**. Once in contact with the very acidic environment of the stomach (pH< 5.0), a peptide fragment is cleaved from the pepsinogen molecule yielding the active protease, **pepsin**. Pepsin can then activate more pepsinogen autocatalytically. In addition to pepsinogen, other zymogens are secreted which yield pepsins B, C and D.

Pepsins (pH optima of 2.5) hydrolyze ingested proteins at sites involving aromatic amino acids, leucine and acidic amino acids. Because of the relatively short residence time of the stomach contents, digestion is limited. The partially digested, relatively large polypeptides then enter the duodenum of the small intestine.

B. Intestinal Digestion

1. The pancreas secretes several proenzymes into the duodenum along with a slightly alkaline fluid buffered to the pH optima of the active forms. The proenzymes include **trypsinogen, chymotrypsinogens, procarboxypeptidases** and **proelastase**.

2. Activations: Upon stimulation by the entrance of food into the intestine the intestinal mucosa secretes the enzyme **enterokinase** which acts on trypsinogen converting it to trypsin. Trypsin in turn activates more trypsinogen and the other proenzymes.

$$\text{Trypsinogen} \xrightarrow{\text{enterokinase}} \text{trypsin + peptide}$$

$$\text{Chymotrypsinogens} \xrightarrow{\text{trypsin}} \text{chymotrypsins + peptides}$$

$$\text{Procarboxypeptidases} \xrightarrow{\text{trypsin}} \text{carboxypeptidase + peptides}$$

$$\text{Proelastase} \xrightarrow{\text{trypsin}} \text{elastase + peptide}$$

3. The following brush border enzymes are secreted into the intestinal lumen but also work intracellularly:

Aminopeptidases- broad specificity; hydrolyze N-terminal amino acids.
Dipeptidases- hydrolyze dipeptides such as glycylglycine.
Prolinase- hydrolyzes peptides containing proline at the N-terminus.

VI. AMINO ACID ABSORPTION

Amino acid absorption is very rapid in the small intestine and is carried out by active transport mechanisms and is, therefore, an energy-expending process. Several specific transport mechanisms have been identified involving different classes of amino acids. These include those for:

1. small neutral amino acids
2. large neutral amino acids
3. basic amino acids
4. acidic amino acids
5. proline

Amino acids of the same class compete with one another for absorption sites.

The Gamma-Glutamyl Cycle

In addition to the transport mechanisms listed above, a general absorption mechanism involving all amino acids with a free amino group has been proposed, referred to as the **gamma-glutamyl cycle.** It provides a role for glutathione and explains the presence of 5-oxyproline in the urine. For every amino acid transported across the membrane, three ATP's are consumed during the regeneration of glutathione.

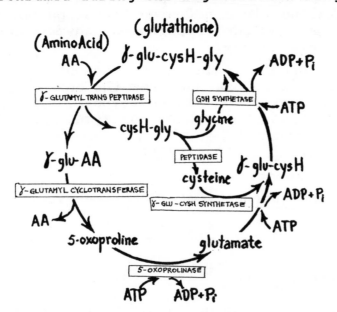

In addition to amino acids, some small peptides are absorbed directly into the blood without hydrolysis. Very rarely, whole proteins are absorbed intact. The major route of entry into the blood is via the portal vein. Once in the blood amino acids are rapidly absorbed into cells. Liver and kidney take up the largest fraction. A blood-brain barrier exists for some amino acids, especially for glutamic acid.

VII. AMINO ACID DEGRADATION

In the breakdown of amino acids the first task the cell must accomplish is the removal of the alpha-amino groups. The two major mechanisms by which this is accomplished are **transamination** and **oxidative deamination.**

A. Transamination

Twelve amino acids can undergo transamination. These are ala, arg, asN, asp, cys, ile, leu, lys, phe, trp, tyr, and val. (Mnemonic: **VAL AT CAPITAL**).

1. The general reaction involved is:

$$\underset{AA_1}{\overset{NH_2}{\underset{|}{R-CH-COOH}}} + \underset{}{\overset{O}{\underset{\|}{R'-C-COOH}}} \longrightarrow \underset{}{\overset{NH_2}{\underset{|}{R'-CH-COOH}}} + \underset{}{\overset{O}{\underset{\|}{R-C-COOH}}}$$

The enzymes involved are called **transaminases** or **aminotransferases**. The reactions catalyzed by transaminases are freely reversible with equilibrium constants approaching 1.0. They can be found both in the mitochondria and the cytosol. The transfer of amino groups from most amino acids to α-ketoglutarate to form glutamate takes place in the cytosol. The glutamate formed can then enter the mitochondria via a special transport mechanism where it can undergo oxidative deamination or else form aspartic acid which can then reenter the cytoplasm.

2. Mechanism of transamination: All transaminases share a common reaction mechanism and use the same cofactor, **pyridoxal phosphate** (**PLP**), which is derived from the vitamin, **pyridoxine** (**vitamin B₆**). PLP is covalently bound to the enzyme via a Schiff's base linkage to an epsilon amino group of a specific lysine located in the active site. During transamination, a Schiff's base forms between the amino acids and PLP.

3. Details of the reaction:

   a. First Stage:

b. Second stage:

$$R_2-\underset{\underset{O}{\|}}{C}-COOH \quad + \quad \underset{\underset{\text{(FROM LAST STAGE)}}{PMP-E}}{H_2N-CH_2-E} \quad \underset{H_2O}{\overset{H_2O}{\rightleftharpoons}} \quad R_2-\underset{\underset{N-CH_2-E}{\|}}{C}-COOH$$

$\underline{\alpha KETOACID_2}$ $\underline{KETIMINE}$

$$R_2-\underset{\underset{NH_2}{|}}{CH_2}-COOH \quad + \quad \underset{PLP-E}{O=\underset{\underset{}{|}}{\overset{H}{C}}-E} \quad \underset{H_2O}{\overset{H_2O}{\rightleftharpoons}} \quad R_2-\underset{\underset{N\neq CH-E}{|}}{CH}-COOH$$

$\underline{AA_2}$ $\underline{ALDIMINE}$

B. Oxidative deamination

1. L-glutamate is the ultimate product from the majority of the transaminations that occur. Glutamate then enters the mitochondria where the following reactions occur.

$$\text{L-glutamate} + NAD^+ + H_2O \longrightarrow \alpha\text{-ketoglutarate} + NH_4^+ + NADH$$

The reaction is catalyzed by **glutamic dehydrogenase** found in both the mitochondria and cytoplasm. The enzyme can use both NAD^+ and $NADP^+$ but NAD^+ is preferred in the catabolic direction. The NADH formed within the mitochondria can enter the electron transport system and yield 3 ATP.

2. <u>Glutamic dehydrogenase</u> is a very complex enzyme made up of six identical 56,000 dalton subunits. It is regulated allosterically by a number of effectors.

      Inhibitors- ATP, GTP, NADH
      Activators- ADP, GDP, specific amino acids
      Hormones- thyroxine and steroid hormones

C. Amino acid oxidases

1. L-amino acid oxidase is found in liver, kidney and snake venom. It is involved primarily in the deamination of lysine:

$$\text{L-amino acid} + H_2O + \underline{E}\text{-FMN} \longrightarrow \alpha\text{-ketoacid} + NH_3 + \underline{E}\text{-FMNH}_2$$

$$(\underline{E} = \text{oxidase molecule})$$

2. D-amino acid oxidase:

$$\text{D-amino acid} + H_2O + \underline{E}\text{-FAD} \longrightarrow \alpha\text{-ketoacid} + NH_3 + \underline{E}\text{-FADH}_2$$

The flavin nucleotides (FMN and FAD) are derived from **riboflavin** (**vitamin B₂**). The regeneration of the flavin nucleotides is carried out utilizing molecular oxygen yielding H_2O_2 which is rapidly metabolized by **catalase**.

VIII. THE UREA CYCLE (KREBS-HENSELEIT CYCLE)

Ammonia is considered to be toxic to cells of the CNS in higher organisms and must be kept at low concentrations. In many aquatic animals like fish and tadpoles, NH3 diffuses directly into the surrounding water through gills. These animals are called **ammonotelic**. Others such as birds and reptiles excrete the ammonia as uric acid (uricotelic). Man and most other terrestrial vertebrates excrete NH₃ as urea (ureotelic).

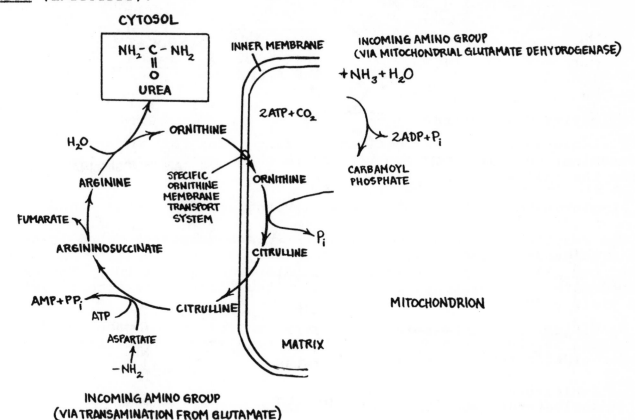

Detailed steps are as follows:

A. Carbamoyl phosphate synthetase (N-acetylglutamate acts as an allosteric activator):

$$2\ ATP + CO_2 + NH_3 + H_2O \longrightarrow H_2N\text{-}\overset{O}{\overset{\|}{C}}\text{-}O\text{-}PO_3 + 2\ ADP + P_1$$
Carbamoyl
phosphate

B. <u>Ornithine carbamoyl transferase</u>

C. <u>Argininosuccinate synthetase</u>

D. <u>Argininosuccinate lyase</u>

E. <u>Arginase</u>

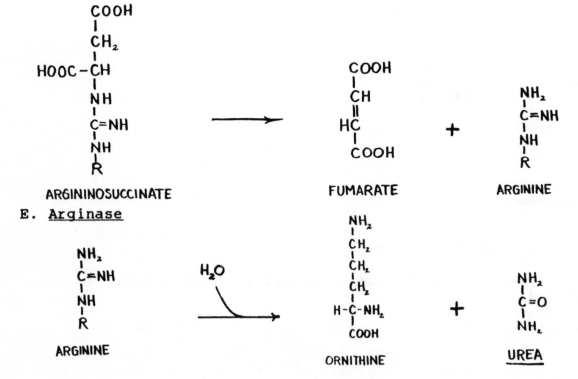

Overall Reaction:

$$2NH_3 + CO_2 + 3ATP + 3H_2O \longrightarrow Urea + 2ADP + AMP + 2P_i + PP_i$$

The urea cycle is intimately tied to the citric acid cycle by way of fumarate and aspartate. One N is derived from NH_3 and one from aspartate. Each of the enzymes of the urea cycle have been identified with specific genetic defects. Upon ingestion of proteins such patients show hyperammonemia, lethargy, vomiting and other signs of CNS disturbance.

IX. EXCRETION OF FREE AMMONIA

In addition to the excretion of urea, **NH₃** can be removed from cells by transferring it to the gamma carboxyl of glutamate to form **glutamine**. Glutamine is non-toxic and can pass through the blood-brain barrier.

$$Glutamate + ATP + NH_3 \xrightarrow{\text{glutamine synthetase}} glutamine + ADP + P_i$$

Glutamine is transported in the blood to the kidneys where it encounters in the tubules the enzyme, **glutaminase**. Glutaminase hydrolyzes the glutamine to glutamate and NH_3. The glutamate is reabsorbed by the tubules and the ammonia is excreted in the urine.

X. DEGRADATION OF THE CARBON SKELETONS

After removal of the α-amino groups, the carbon skeletons of the 20 amino acids undergo a series of reactions that result in products that are members of the glycolytic pathway, the citric acid cycle or keto-acids. There are only seven of these: pyruvate, acetyl CoA, acetoacetyl CoA, α-ketoglutarate, succinyl CoA, fumarate and oxaloacetate.

Table 5.1 outlines the catabolism of the 20 amino acids.

XI. ONE-CARBON FRAGMENT METABOLISM

In the degradation and synthesis of amino acids (and many other compounds), there is often a need to remove, add or rearrange 1-C units. With the exception of the decarboxylases which remove 1-C units as CO_2 or HCO_3^-, all 1-carbon transfers require the participation of cofactors. The precursors of these cofactors must be supplied by the diet, either as vitamins or as essential amino acids.

A. Tetrahydrofolate (THFA) (tetrahydropteroylglutamate)

1. Structure: This compound is the most versatile of the carriers of 1-C fragments and carries them at several levels of oxidation, from the most reduced, methyl, to the most oxidized, methenyl (Figure 5.1):

Amino Acid	Products	Number of Enzymatic Steps	Cofactors	Glycogenic or Lipogenic
Alanine	Pyruvate	1	PLP	G
Glycine	Pyruvate	2	N^5,N^{10}-methylene THFA	G
Serine	Pyruvate	1		G
Cysteine	Pyruvate	2	PLP,NADH	G
Threonine	Pyruvate			G
. . . .				
Aspartic Acid	Oxaloacetate	1	PLP	G
Asparagine	Oxaloacetate	2	PLP	G
. . . .				
Histidine	α-ketoglutarate	5	THFA,PLP	G
Glutamic acid	α-ketoglutarate	1	PLP	G
Glutamine	α-ketoglutarate	2	PLP	G
Arginine	α-ketoglutarate	4	PLP,NAD	G
Proline	α-ketoglutarate	4	O_2, PLP	G
. . . .				
Methionine	succinyl CoA	9	ATP,CoA,NAD,biotin,Vit B_{12}	G
Valine	succinyl CoA	10	PLP,NAD,CoA,Vit B_{12}	G
Isoleucine	succinyl CoA	9	PLP,NAD,CoA,FAD,biotin, Vit B_{12}	G,L
. . . .				
Leucine	acetyl CoA, acetoacetyl CoA		thiamine PP,lipoic acid, biotin,Vit B_{12},PLP, CoA,NAD,FAD	L
Phenylalanine	acetoacetyl CoA, fumarate	7	O_2,NADPH, tetrahydrobiopterin	G,L
Tyrosine	acetoacetyl CoA fumarate	6	O_2,NADPH, tetrahydrobiopterin	G,L
Tryptophan	acetoacetyl CoA alanine	9	O_2,NADPH,NAD	G,L
Lysine	acetoacetyl CoA	9	NADPH,NAD,NADP,PLP,CoA,FAD	G,L

Table 5.1. Catabolism of the 20 Amino Acids

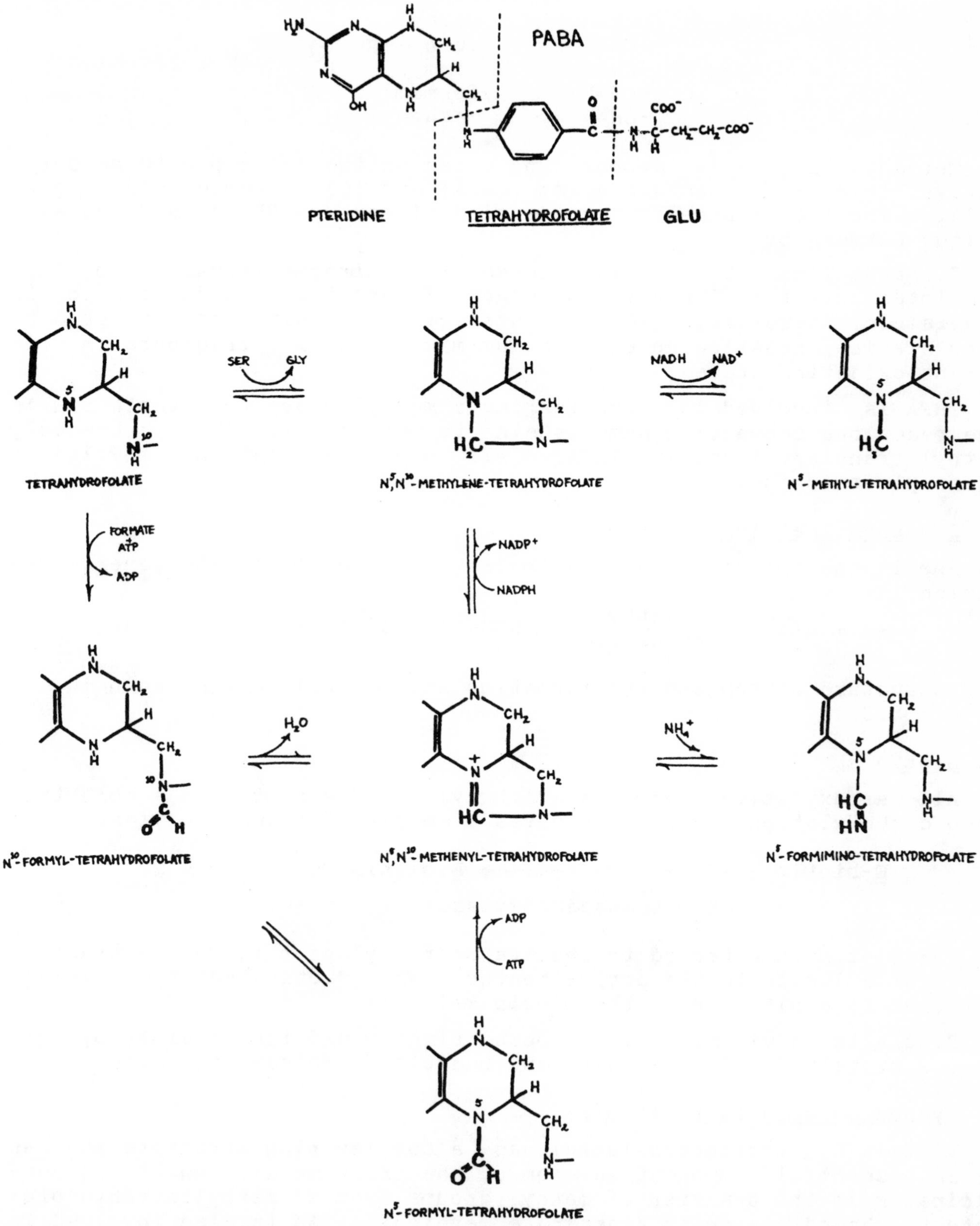

Figure 5.1. Structure of Tetrahydrofolic Acid and the Metabolism of the Various C-1-Carrying Forms.

$-CH_3$	methyl
$-CH_2-$	methylene
$-CH=O$	formyl
$-CH=NH$	formamino
$-CH=$	methenyl

Tetrahydrofolate is composed of three units: (a) a pteridine derivative, (b) p-aminobenzoic acid (PABA) and (c) glutamic acid (Figure 5.1). The 1-C fragments are carried either on N^5 or N^{10} or as a bridge between both.

2. Metabolism: The several forms of 1-C fragments carried by THFA are interconvertible one to the other (Figure 5.1). Each reaction is reversible, therefore, 1-C fragments can be donated or accepted by each species, creating an equilibrium mixture of 1-C fragments in different oxidation states.

THFA is involved in transferring a methyl group to Vitamin B_{12} in the reactions converting homocysteine to methionine. This is the only methyl transfer involving THFA. All other methylations involve S-adenosylmethionine.

B. S-Adenosylmethionine (SAM)

SAM is the cofactor involved in transmethylations. The general reaction is:

$$\text{SAM} + \text{R(acc)} \xrightarrow{\text{methyl transferase}} \text{RCH}_3 + \text{S-adenosylhomocysteine}$$

The structure of SAM and its formation and metabolism are shown in Figure 5.2.

C. Biotin

All carboxylations via transcarboxylases use **biotin** as a cofactor. Each carboxylation reaction involves a specific transcarboxylase.

$$\underline{\text{E}}\text{-biotin} + \text{ATP} + \text{HCO}_3^- \longrightarrow \underline{\text{E}}\text{-biotin-CO}_2 + \text{ADP} + \text{P}_1$$

$$\underline{\text{E}} = \text{transcarboxylase}$$

The biotin is attached to the transcarboxylase through the ε-amino group of a lysine in the active center. When "activated" the CO_2 is attached to a nitrogen of the biotin molecule.

Biotin is a vitamin. It is bound tightly and irreversibly by avidin, a protein found in significant quantities in raw egg white.

D. Cyanocobalamin (Vitamin B12)

Vitamin B_{12} (cyanocobalamin) has a complex ring structure and has as an essential component an atom of the trace metal, **cobalt**. It participates in the transfer of methyl groups from N^5-methyltetrahydrofolate to homocysteine to regenerate methionine. It is also involved in certain one-carbon rearrangement reactions as in the conversion of

1. Synthesis of SAM from Methionine:

METHIONINE + ATP $\longrightarrow$ P_i + PP_i + **S-ADENOSYLMETHIONINE**

2. Transfer of Methyl Group:

S-ADENOSYLMETHIONINE $\xrightarrow{R \quad R-CH_3}$ **S-ADENOSYL-HOMOCYSTEINE** $\xrightarrow{H_2O \quad ADENOSINE}$ **HOMOCYSTEINE**

3. Re-synthesis of Methionine:

HOMOCYSTEINE + **N⁵-METHYLTETRAHYDROFOLATE** $\xrightarrow{\text{VITAMIN B 12}}$ **METHIONINE** + **TETRAHYDROFOLATE**

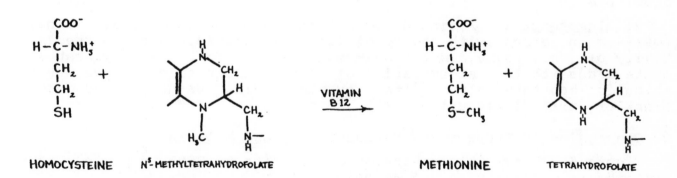

Figure 5.2. S-Adenosylmethionine: Structure, Formation and Metabolism.

methylmalonyl CoA to succinyl CoA in the latter stages of the degradation of Met, Val and Ile. An inability to obtain enough Vitamin B_{12} in the diet or an inability to absorb it in the intestinal tract leads to the condition of **pernicious anemia**.

XII. METABOLISM OF PHENYLALANINE AND TYROSINE

Several genetic defects exist involving the enzymes of phenylalanine and tyrosine degradation. Tyrosine is also the precursor to several neurohormones and to melanin. Therefore, the metabolism of these amino acids will be covered in slightly more detail.

A. Degradation of Phenylalanine and Tyrosine

The scheme for the degradation of phenylalanine and tyrosine is shown in Figure 5.3.

1. Phenylketonuria: The most common genetic disturbance in the metabolism of these amino acids occurs at the first step, the oxidation of phenylalanine to tyrosine by phenylalanine hydroxylase (phenylalanine 4-monooxygenase). The absence of this enzyme or a defect in its functioning caused by defects in other components of the system leads to the accumulation of **phenylpyruvate** causing the condition known as **phenylketonuria**. This condition has a frequency of about 1/10,000 and if untreated is characterized by mental retardation, CNS damage and hypopigmentation. If diagnosed early, some affected children may be spared from the major damaging effects of phenylpyruvate accumulation by placing them on a diet low in phenylalanine, substituting the phenylalanine with tyrosine.

An essential cofactor for phenylalanine hydroxylase is **tetrahydrobiopterin** which is also oxidized, during the oxidation of phenylalanine to tyrosine, to **dihydrobiopterin**. The dihydro form must be reduced back to the tetrahydro form to sustain the overall conversion reaction. The enzyme carrying out this reduction uses NADPH as a cofactor. Some cases of phenylketonuria are the result of defects in this enzymatic step or in steps related to the synthesis of tetrahydrobiopterin.

2. Alcaptonuria: A second defect in tyrosine catabolism has been observed in persons with the condition **alcaptonuria**. A diminished activity of the enzyme converting **homogentisic acid** to 4-maleylacetoacetate leads to the accumulation of homogentisic acid in the urine which is spontaneously oxidized by atmospheric oxygen to produce a black or dark-colored urine. This condition is relatively benign.

B. Conversion of Tyrosine to Neurohormones and Melanin

Figure 5.4 shows the conversion of tyrosine to DOPA, dopamine, norepinephrine, epinephrine and melanin.

DOPA (dihydroxyphenylalanine) is a key intermediate in these pathways. Genetic defects in the conversion of DOPA to **melanin** causes **albinism**, a condition of very severe hypopigmentation. **Dopamine** and **norepinephrine** are neurotransmitters. **Epinephrine** is synthesized pri-

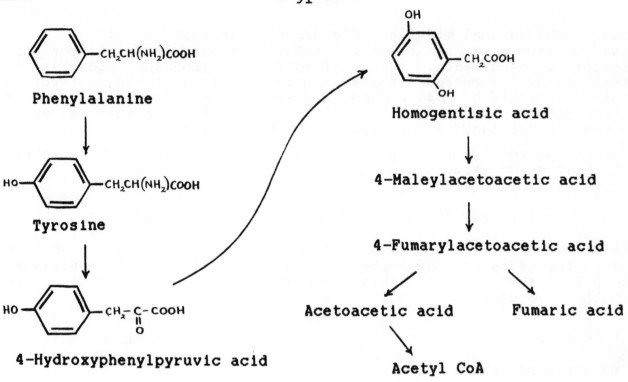

Figure 5.3. Degradation of Phenylalanine and Tyrosine

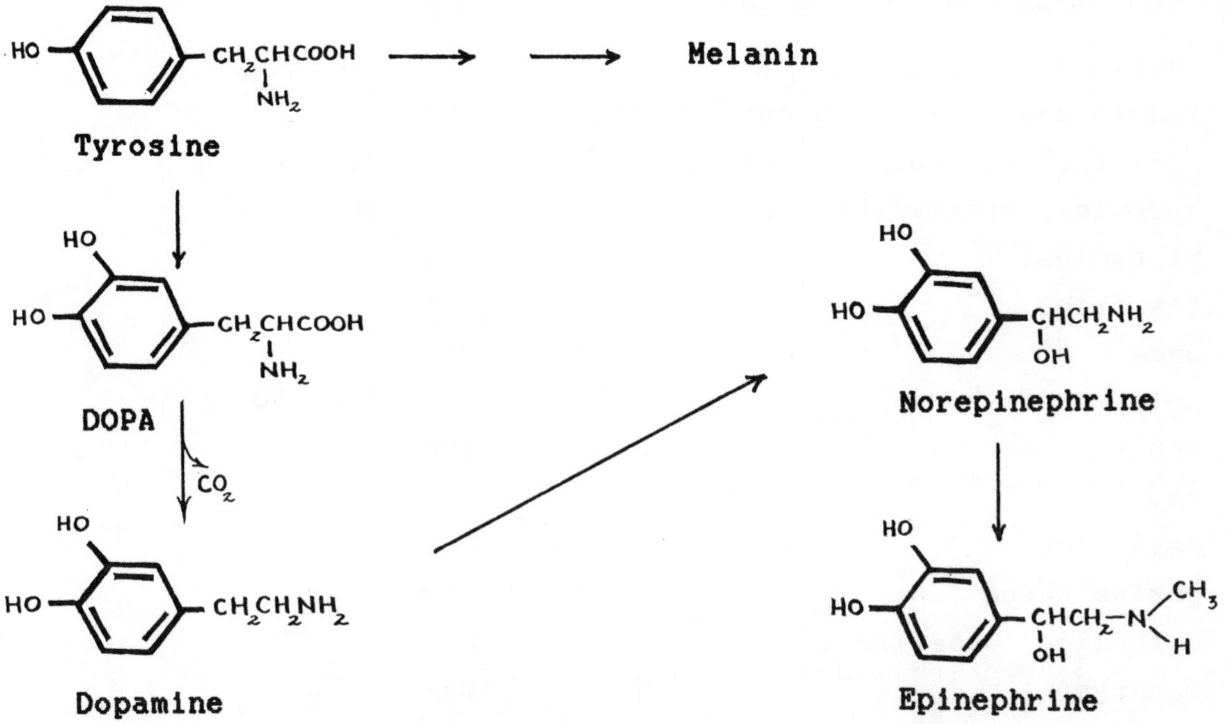

Figure 5.4. Metabolism of Tyrosine

marily in the adrenal medulla. Disturbances in the functioning of CNS pathways using dopamine as a transmitter have been linked to the condition of **schizophrenia**. A lack of sufficient dopamine production in certain brain structures like the *substantia nigra* or the destruction of this structure by toxic compounds leads to **Parkinson's Disease** or the **Parkinsonian syndrome**. Administration of large quantities of L-DOPA can reverse many of the symptoms in some cases.

Norepinephrine is found primarily at nerve endings of the adrenergic (sympathetic) nervous system.

XIII. GENERAL PRECURSOR FUNCTIONS OF AMINO ACIDS

In addition to serving as precursors to proteins, amino acids also act as precursors to many other compounds of biological importance. Table 5.2 lists a few of these compounds and the amino acid(s) from which they are derived.

Compound	Amino Acid Precursor(s)
Neurotransmitters	
many amino acids serve directly as neurotransmitters	
dopamine, epinephrine,norepinephrine	Phe, Try
serotonin (5-hydroxytryptamine)	Trp
GABA (γ-aminobutyric acid)	Glu
Miscellaneous compounds	
indole acetic acid (plant hormone)	Trp
creatine and phosphocreatine	Met, Gly, Arg
spermine, spermidine	Met, Arg (Orn)
histamine	His
thyroxine	Tyr
heme	Gly
polypeptide hormones	all 20 amino acids
NAD	Trp
taurine	Cys
carnitine	Lys
purine bases	Gly
carnosine, anserine	His
Coenzyme A	Cys

Table 5.2. Important Derivatives of Amino Acids

XIV. REVIEW QUESTIONS ON AMINO ACID METABOLISM

DIRECTIONS: Each of the questions or incomplete statements below is followed by five suggested answers or completions. Select the one that is BEST in each case and fill in the corresponding space on the answer sheet.

1. Upon degradation, serine, alanine and cysteine are likely to be converted to:

A. α-ketoglutarate
B. pyruvate
C. fumarate
D. succinate
E. none of the above

2. Pathways for the synthesis of pyrimidines, urea, and citrulline have in common a requirement for:

A. acetate
B. carbamyl phosphate
C. tetrahydrofolate
D. propionate
E. pyruvate

3. The primary site of urea synthesis is in the:

A. kidney
B. skeletal muscles
C. liver
D. small intestine
E. brain

4. The major non-protein nitrogenous component of blood is:

A. urea
B. ammonia
C. a purine
D. an amino acid
E. uric acid

5. The major source of ammonia ($NH4^+$) in fresh urine is from:

A. the hydrolysis of glutamine by glutaminase
B. the hydrolysis of urea by urease.
C. the oxidation of amino acids by L-amino acid oxidase
D. the oxidation of amines by an amine oxidase
E. deamination of aspartate by aspartate ammonia lyase

6. Which of the following hormones affects the urinary concentration of urea?

A. insulin
B. epinephrine
C. vasopressin
D. pancreozymin
E. thyroxine

7. Which of the following biosyntheses does not involve the participation of folic acid?

A. methionine
B. serine
C. purines
D. epinephrine
E. tyrosine (mammalian)

8. The coenzyme involved in transaminations and many other amino acid transformations is derived from:

A. niacin
B. pyridoxine (vitamin B_6)
C. flavins
D. thiamine
E. vitamin B_{12}

9. Which of the following is not a donor of methyl groups?

A. methionine
B. creatine
C. S-adenosylmethionine
D. N^5,N^{10}-methylenetetrahydrofolate
E. betaine

10. Compounds important to neural function resulting, in full or part, from amino acid decarboxylations include:

A. norepinephrine
B. gamma-amino butyrate
C. serotonin
D. acetyl choline
E. all of the above

11. The phosphorylated form of creatine is:

A. a source of high energy phosphate for ATP formation in muscle
B. a component of the urea cycle
C. excreted by the kidney
D. an important hormone
E. none of the above

12. Which of the following is not involved in the biosynthesis of creatine phosphate?

A. pyridoxal phosphate
B. ATP
C. methionine
D. glycine
E. arginine

13. Dietary methionine can be replaced entirely by which one of the following compounds?

A. threonine
B. homocysteine
C. folic acid
D. cysteine
E. creatine

14. Which one of the following amino acids is non-essential as a human nutrient?

A. lysine
B. phenylalanine
C. valine
D. threonine
E. proline

15. The tripeptide, glutathione, does NOT contain:

A. glycine
B. cysteine
C. glutamate
D. threonine

16. Beta-alanine is a constituent of:

A. glutathione
B. carnosine
C. flavin mononucleotide
D. NAD
E. biotin

17. A positive nitrogen balance is likely to be found in:

A. a growing child
B. a healthy adult
C. a senescent adult
D. a child on a lysine-deficient diet
E. an adult on a phenylalanine deficient diet

18. In Maple Syrup Urine disease (now called branched chain aminoaciduria), the defective metabolic step involves:

A. an oxidative decarboxylation
B. an amino acid transaminase
C. a methionine deficiency in the diet
D. an amino acid hydroxylase
E. amino group fixation to carbon skeletons

19. In the genetic defect, homocystinuria, the defective enzyme is thought to be:

A. methionine demethylase
B. cystathionine synthetase
C. cystathionase
D. S-adenosylhomocysteine hydrolase
E. a kidney enzyme involved in amino acid transport

20. Which one of the following amino acids is solely ketogenic?

A. isoleucine
B. phenylalanine
C. leucine
D. proline
E. tryptophan

21. The intravenous injection of a labeled, radioactive amino acid to a mature animal will lead to:

A. no incorporation of labeled amino acid into protein
B. incorporation of equivalent amounts of labeled amino acid into all the proteins of the body
C. incorporation of different amounts of labeled amino acid into various proteins of the body
D. rapid and complete excretion of the labeled amino acid
E. excretion of an equal amount of unlabeled amino acid

22. Serine and cysteine may enter the citric acid cycle as acetyl CoA after conversion to:

A. succinyl CoA
B. pyruvate
C. oxaloacetate
D. glyoxylic acid
E. propionate

23. Which amino acid is both ketogenic and glycogenic?

A. leucine
B. valine
C. arginine
D. histidine
E. lysine

24. Which amino acid is directly involved in transfer of 1-carbon fragments?

A. methionine
B. tryptophan
C. proline
D. tyrosine
E. threonine

25. Tryptophan is not involved in the biosynthesis of:

A. niacin
B. serotonin
C. norepinephrine
D. melatonin
E. indoles

26. Which enzymatic step is not involved in the biosynthesis of epinephrine?

A. an amino oxidation
B. a methylation
C. an aliphatic hydroxylation
D. an aromatic hydroxylation
E. a decarboxylation

27. Each of the following conversions occurs in humans EXCEPT:

A. serine to cysteine
B. phenylalanine to tyrosine
C. glutamate to proline
D. oxaloacetate to lysine
E. homocysteine to methionine

28. Animal proteins are generally nutritionally superior to plant proteins because:

A. More nutritionally valuable carbohydrates are found in animal glycoproteins
B. The average content of glycine and serine in animal protein is greater
C. More B vitamins are adsorbed to the surface of animal proteins
D. The average content of tryptophan and lysine in animal protein is greater

29. If a mature animal in nitrogen balance is placed on a diet deficient only in phenylalanine, which of the following conditions is most likely to occur?

A. Nitrogen balance will become negative and remain that way so long as the deficiency exists.
B. Nitrogen balance will become negative temporarily, but the individual will adapt and nitrogen balance will gradually return to zero.
C. Nitrogen intake will continue to equal nitrogen excretion (balance= 0).
D. Nitrogen balance will become positive and remain that way so long as the deficiency exists.
E. Nitrogen balance will become positive temporarily, but the individual will adapt and nitrogen balance will gradually return to zero.

30. Which one of the following must be supplied to the normal human adult in order to maintain nitrogen balance?

A. threonine
B. alanine
C. lysine
D. tyrosine
E. serine

31. Which of the following enzymes is most important in the process of dietary protein digestion (and especially zymogen activation)?

A. enterokinase
B. prolidase
C. carboxypeptidase B
D. pepsin
E. chymotrypsin

32. Which of the following enzyme pairs is involved in the conversion of amino acid nitrogen into two compounds that directly provide the urea nitrogen?

A. glutamic-oxaloacetic transaminase and diamine oxidase
B. L-amino acid oxidase and racemase
C. serine dehydratase and glutamic dehydrogenase
D. carbamyl phosphate synthetase and glutamic-oxaloacetic transaminase
E. glutamine synthetase and glutaminase

33. The carbon skeletons of serine, glycine and cysteine are readily fed into the citric acid cycle and serve as sources of energy. The non-nitrogenous intermediate through which they are converted to acetyl CoA is:

A. α-ketoglutarate
B. pyruvate
C. fumarate
D. oxaloacetate
E. isocitrate

34. Conversion of ornithine to citrulline is a step in the synthesis of:

A. arginine
B. cysteine from methionine
C. tyrosine form glucose
D. urea from NH_3
E. A and D are correct

35. Substances that can contribute carbon atoms to the net synthesis of glycogen are also convertible to:

A. glucose
B. alanine
C. glycerol
D. phenylalanine
E. 1,2, and 3 are correct

36. After glucose labeled in the C-1 position with ^{14}C was administered to a mouse, the animal's proteins became labeled. Which of the following amino acids, derived from these proteins, would NOT be labeled?

A. proline
B. leucine
C. aspartate
D. cystine
E. alanine

37. Enzymes acting in the urea cycle include:

A. glutaminase
B. urease
C. acetylornithinase
D. arginase
E. none of the above

38. A direct donor of a nitrogen atom to urea is:

A. ornithine
B. methionine
C. aspartic acid
D. glutamic acid
E. creatinine

39. Which of the following would most strongly initiate gastrin release upon ingestion?

A. ethanol
B. fats
C. glucose
D. protein
E. DNA

40. Zymogen molecules are converted to active enzymes as a result of:

A. enzymatic phosphorylation
B. activation by ATP
C. enzyme adenylate formation
D. limited proteolysis

41. The biosynthesis of glucose from aspartate involves:

A. dephosphorylation
B. transamination
C. aldol condensation
D. carboxylation
E. all of the above

42. Hormones derived from amino acid precursors include those of the:

A. pancreas
B. thyroid gland
C. adrenal medulla
D. parathyroid glands
E. all of the above

DIRECTIONS: Each group of items below consists of lettered headings followed by a set of numbered words or phrases. For each numbered word or phrase, select the ONE heading that is most closely associated with it and fill in the corresponding space on the answer sheet. Each heading may be used once, more than once, or not at all.

Questions 43-45:

   A. α-ketoglutaric acid
   B. glutamic acid
   C. aspartic acid
   D. acetyl CoA
   E. pyruvic acid

43. Can be deaminated enzymatically to form oxaloacetic acid

44. Is formed directly from alanine by transamination

45. Deamination directly yields α-ketoglutaric acid

Questions 46-54: Which of the following compounds are most directly involved as precursors in the biosynthesis of the metabolites in questions 46-54?

   A. tyrosine
   B. tryptophan
   C. serine
   D. glutamic acid
   E. glycine

46. porphyrins

47. γ-aminobutyric acid

48. serotonin

49. kynurenine

50. epinephrine

51. acetylcholine

52. thyroxine

53. niacin

54. creatine

Questions 55-58:

   A. enterokinase
   B. chymotrypsinogen
   C. carboxypeptidase
   D. pepsin
   E. trypsin

55. An intestinal enzyme which initiates zymogen activation.

56. An exopeptidase

57. A protease with a pH optimum less than 3.0.

58. A precursor to a proteolytic enzyme

Questions 59-60:

   A. glucose formation in the liver from glycerol
   B. glucose formation in the liver from alanine
   C. both A and B
   D. neither A nor B

59. NADH required

60. ATP required

Questions 61-63:

   A. urea cycle
   B. activation of protease precursors
   C. decarboxylation of oxaloacetate
   D. fixation of free ammonia
   E. generation of free amino acids from protein

61. Enterokinase

62. Arginase

63. Exopeptidase

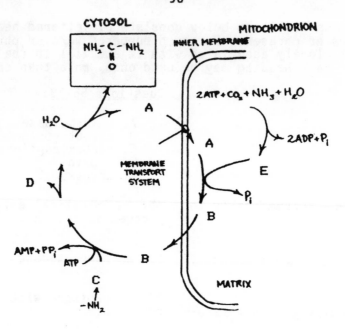

Questions 64-68: In the diagram above, the various numbered components of the urea cycle are:

64. ornithine

65. component of the citric acid cycle

66. carbamoyl phosphate

67. citrulline

68. aspartate

Questions 69-73: Match the coenzymes listed below with the processes in questions 69-73.

    A. pyridoxal phosphate
    B. tetrahydrofolate
    C. adenosine 5'-triphosphate
    D. a flavin nucleotide
    E. NADH

69. The interconversion of serine and glycine

70. Transaminations

71. The formation of S-adenosyl methionine

72. The conversion of δ-pyrroline-5-carboxylate to proline

73. The oxidation of amines

Questions 74-77: Aromatic amino acid metabolism involves the following pathway:

    phenylalanine

    A ↓

    tyrosine $\xrightarrow{B}$ melanin

    C ↓

    p-hydroxyphenylpyruvate

    D ↓

    homogentisic acid

    E ↓

    maleoyl acetoacetic acid

Identify the genetic or nutritional condition listed below with an enzymatic block (lettered in diagram above).

74. tyrosinosis

75. albinism

76. phenylketonuria

77. alkaptonuria

XV. ANSWERS TO QUESTIONS ON AMINO ACID METABOLISM

1. B	21. C	41. E	61. B
2. B	22. B	42. E	62. A
3. C	23. E	43. C	63. E
4. A	24. A	44. E	64. A
5. A	25. C	45. B	65. D
6. C	26. A	46. E	66. E
7. D	27. D	47. D	67. B
8. B	28. D	48. B	68. C
9. A	29. A	49. B	69. B
10. E	30. C	50. A	70. A
11. A	31. A	51. C	71. C
12. A	32. D	52. A	72. E
13. B	33. B	53. B	73. D
14. E	34. E	54. E	74. C
15. D	35. E	55. A	75. B
16. B	36. B	56. C	76. A
17. A	37. D	57. D	77. E
18. A	38. C	58. B	
19. B	39. A	59. A	
20. C	40. D	60. C	

6. PORPHYRINS

Thomas Briggs

I. STRUCTURE AND CHEMISTRY

Porphyrins: cyclic tetrapyrroles found in **all aerobic cells**. With a system of conjugated double bonds, they absorb light strongly and are intensely colored, as in blood.

PROTOPORPHYRIN IX:

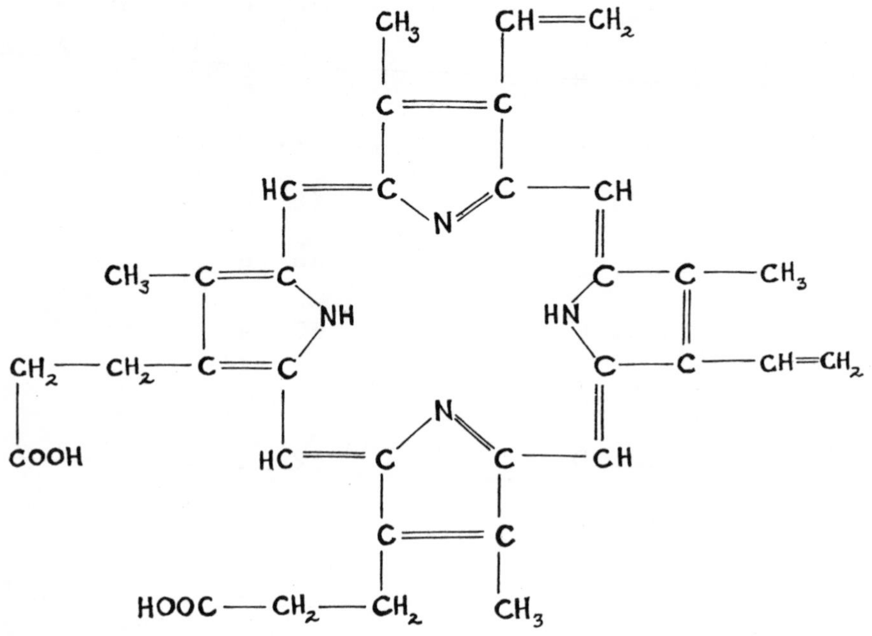

Porphyrins vary in the type and sequence of side-chains on the tetrapyrrole ring. Many are derivatives of **Protoporphyrin IX**, which has the particular sequence shown above. Biologically active forms usually have a **metal ion** in the center, are associated with a **protein**, and take part in **enzymic activities**.

Heme: iron protoporphyrin IX. It occurs in:
    hemoglobin and **myoglobin** (O_2 transport and storage)
    cytochromes (electron transport)
    catalase (simple breakdown of H_2O_2)
    peroxidases (use H_2O_2 to oxidize an organic substrate).

Tetrapyrroles (not heme) are also part of:
    chlorophyll
    vitamin B_{12}.

II. SYNTHESIS

$$\underset{\text{SUCCINYL CoA}}{\overset{\text{COOH} \quad \text{O}}{\text{CH}_2\text{-CH}_2\text{-C-SCoA}}} + \underset{\text{GLYCINE}}{\overset{\text{COOH}}{\text{CH}_2\text{-NH}_2}} \xrightarrow[\substack{\text{ALA} \\ \text{SYNTHASE}}]{\substack{\text{PYRIDOXAL} \\ \text{PHOSPHATE}}} \underset{\substack{\delta\text{-AMINO-LEVULINIC ACID} \\ \text{(ALA)}}}{\overset{\text{COOH} \quad \text{O}}{\text{CH}_2\text{-CH}_2\text{-C-CH}_2\text{-NH}_2}} + CO_2 + CoASH$$

ALA synthase is the first enzyme in the chain of heme synthesis. An important control point, it is rate-limiting and inhibited by the end product of the sequence, heme (or derivatives such as hemin [ferriheme]). Heme (and hemoglobin) synthesis is stimulated by **erythropoietin**, a glycoprotein.

Two ALA are condensed to **porphobilinogen** (PBG), a monopyrrole. Four of these are joined, under the influence of two enzymes, **PBG deaminase** (older term, Uro III synthetase) and **Uroporphyrinogen III cosynthase** to produce one particular isomeric cyclic tetrapyrrole, a porphyrinogen (**Uroporphyrinogen III**). This is then trimmed through several decarboxylations, giving rise to the specific sequence of methyl, vinyl, and propionic acid side-groups that is to be found in heme. The colorless product is next oxidized with NADPH and molecular oxygen to produce the colored protoporphyrin. Finally, iron is inserted by the enzyme **ferrochelatase**.

Heme synthesis occurs in all aerobic cells, but especially in blood-forming tissues. In man 80% of the body's heme is in **hemoglobin**, which reversibly binds and transports O_2. Carbon monoxide competes strongly with O_2, accounting for its toxicity.

III. METABOLISM

Heme is degraded in the reticuloendothelial system of liver, spleen and bone marrow. The metabolism involves a number of stages:

1. The ring system is opened to form biliverdin which is oxidized to **bilirubin**, a linear tetrapyrrole, an orange-yellow pigment.

2. Bilirubin, (insoluble in water) is transported to the liver as a serum **albumin complex**.

3. Bilirubin is then taken up by liver cells as a complex with a binding protein, **ligandin**.

4. To be excreted, bilirubin must be **conjugated with glucuronic acid**:

Bilirubin + UDP Glucuronic acid $\xrightarrow{\text{glucuronyl transferase}}$ Bilirubin diglucuronide

5. The glucuronide is **excreted in bile**.

Bilirubin is the major **bile pigment** in man. Bacterial metabolism of bilirubin in the bowel leads to formation of a complex mixture of pigments, including **urobilinogen**. Some of these undergo an enterohepatic circulation and may also be found in urine.

IV. DISORDERS

A. Synthesis

These diseases usually take the form of overproduction and excretion of precursors (ALA, PBG) or porphyrin intermediates. Urine may be red. Called **porphyrias**, these conditions are usually inherited, but can sometimes be caused by toxic agents such as lead or some pesticides. They may be accompanied by photosensitivity of the skin, intermittent pain, muscular paralysis and psychic disturbances. ("Mad King George III" was not mad, but probably had one of the heritable porphyrias.) See Chapter 11. **Hematin** is useful in treatment because it inhibits ALA synthase.

B. Metabolism

Problems can occur at many points; build-up of bilirubin in blood and tissues causes a visible yellow color called **jaundice**. A mild form often occurs in the newborn, due to a temporary underdevelopment of the conjugation and excretion process. Severe and prolonged accumulation of unconjugated bilirubin in the newborn is dangerous because unconjugated bilirubin can penetrate nervous tissue, concentrate in brain cells (**kernicterus**), and cause mental retardation.

Area of Defect in Bilirubin Metabolism	Cause	Main Form of Bilirubin Found
overproduction	hemolytic disorders	Unconjugated
uptake, conjugation and secretion by liver	hepatitis, cirrhosis	Variable
biliary transport	blockage of bile duct by gallstone or cancer of pancreas	Conjugated
various	heritable defects in metabolism and transport, see Chapter 11	

V. REVIEW QUESTIONS ON PORPHYRINS

DIRECTIONS: Each of the questions or incomplete statements below is followed by five suggested answers or completions. Select the one that is BEST in each case and fill in the corresponding space on the answer sheet.

1. Which of the following can be converted to bile pigments?

A. cholic acid
B. heme
C. triglycerides
D. cholesterol
E. phospholipids

2. Porphyrins are made from which of the following starting materials?

A. succinic acid and glycine
B. lysine and proline
C. alanine and glutamic acid
D. acetyl CoA and oxaloacetate
E. glycerol and fatty acid

3. Urobilinogen can NOT be derived from

A. hemoglobin
B. catalase
C. flavoprotein
D. cytochrome c
E. peroxidase

4. Urobilinogen is formed in the

A. spleen
B. liver
C. bone marrow
D. bowel
E. kidney

5. Which of the following does NOT contain an iron-porphyrin?

A. carboxyhemoglobin
B. catalase
C. peroxidase
D. myoglobin
E. ferredoxin

6. The function of erythropoietin is to

A. promote synthesis of hemoglobin
B. prevent pernicious anemia
C. prevent thalassemia
D. promote binding of oxygen to hemoglobin
E. regulate reduction of methemoglobin

7. Porphyria is associated with

A. phenylalanine-ammonia lyase deficiency
B. hypoxanthine-guanine phosphoribosyltransferase deficiency
C. overproduction of δ-aminolevulinic acid
D. bilirubin glucuronyl transferase deficiency
E. ALA synthetase deficiency

8. In porphyrin synthesis, the first committed step is

A. condensation of 2 PBG (porphobilinogen)
B. condensation of glycine and succinyl CoA
C. isomerization of isopentenyl-PP to dimethylallyl-PP
D. condensation of 2 ALA (δ-aminolevulinic acid)
E. formation of uroporphyrinogen III

DIRECTIONS: For each of the questions or incomplete statements below, ONE or MORE of the answers or completions is correct. On the answer sheet fill in space

A if only 1, 2, and 3 are correct
B if only 1 and 3 are correct
C if only 2 and 4 are correct
D if only 4 is correct
E if all are correct

FILL IN ONLY ONE SPACE ON YOUR ANSWER SHEET FOR EACH QUESTION

Directions Summarized				
(A) 1,2,3 only	(B) 1,3 only	(C) 2,4 only	(D) 4 only	(E) All are correct

9. Bilirubin diglucuronide is

1. elevated in neonatal hyperbilirubine-mia
2. usually found in the bile duct
3. lipid-soluble
4. water-soluble

10. A consequence of bile duct obstruction is

1. increased conjugated bile pigments in the serum
2. decreased excretion of cholesterol
3. decreased levels of bilirubin glucuronides in the feces
4. increased levels of fecal lipids

11. Catalase

1. is a heme protein
2. is found in peroxisomes
3. breaks down hydrogen peroxide
4. oxidizes organic substrates

12. Unconjugated bilirubin is

1. elevated in neonatal hyperbilirubine-mia
2. usually found in the bile duct
3. lipid-soluble
4. water-soluble

13. Peroxidase

1. is a heme protein
2. is found in peroxisomes
3. breaks down hydrogen peroxide
4. oxidizes organic substrates

VI. ANSWERS TO QUESTIONS ON PORPHYRINS

1. B	5. E	8. B	11. A
2. A	6. A	9. C	12. B
3. C	7. C	10. E	13. E
4. D			

7. LIPIDS

Chi-Sun Wang

I. CLASSIFICATION OF LIPIDS

A. Major Lipid Classes

1. **Fatty Acids:** long-chain monocarboxylic acids.

2. **Acylglycerols:** triacylglycerols, diacylglycerols and monoacyl-glycerols.

3. **Phospholipids:** phosphoglycerides and sphingomyelin.

4. **Glycosphingolipids:** cerebroside (ceramide monosaccharide), cerebroside sulfate, ceramide oligosaccharide and ganglioside. The common structural component of glycosphingolipids is ceramide (N-acylsphingosine). They are similar to sphingomyelin as derivatives of a ceramide, but do not contain phosphorus and the additional nitrogenous base.

5. Other Lipids: **sterols, terpenes, waxes, aliphatic alcohols.**

6. Lipids Combined with Other Classes of Compounds: **lipoproteins.**

B. Derived Lipids

These are molecules, soluble in lipid solvents, that are produced by hydrolysis of natural lipids.

II. FUNCTIONS OF LIPIDS

A. Structural Component

All cellular membranes, including myelin, consist of a **bilayer** which functions as a permeability barrier.

B. Enzyme Cofactors

Several enzymes require lipids for their activity. Phospholipid in blood clotting cascade; coenzyme A, etc.

C. Energy Storage

The major function of triacylglycerols which are largely in **adipose tissue.** These lipids do not require hydration and yield **9 Kcal/g** of energy on complete combustion compared to 4 for carbohydrates and 4 for proteins. This is because the C-H/O ratio is higher.

D. Hormones and Vitamins produced from lipids.

　　1. **Prostaglandins**:　arachidonic　acid　is　the　precursor　for　the biosynthesis of prostaglandins.

　　2. **Steroid** Hormones

　　3. Fat-soluble **Vitamins**:　A, D, E, K.

E. Aid in Fat Digestion:　components of bile.

F. Insulation:　subcutaneous tissue.

III. DIGESTION AND ABSORPTION OF LIPIDS

Most dietary　lipid　is　in the　form of triacylglycerols and is digested　in　the **small　intestine**.　The emulsification of lipid droplets by **bile　salt** allows the increase of the droplet surface area.　Lipase produced by　the pancreas catalyzes the hydrolysis of triacylglycerols to fatty　acids and monoacylglycerols, which form **micelles**.　These micelles also　contain bile salts, cholesterol and fat-soluble vitamins. The micelles　then migrate by diffusion to intestinal mucosa and enter the epithelial cells, where the free fatty acids and monoacylglycerols are re-esterified　to　form　triacylglycerols.　The　triacylglycerol, cholesterol, cholesterol　ester, phospholipid,　and　specific　proteins are assembled　into **chylomicron** particles which subsequently enter the lymphatics for transport to the thoracic duct and the bloodstream.

IV. FATTY ACIDS

A. Chemistry of Fatty Acids

　　1. Some Common Saturated Fatty Acids

Name	Nbr of Carbons	Structure
Acetic	2	CH_3COOH
Propionic	3	CH_3CH_2COOH
Butyric	4	$CH_3(CH_2)_2COOH$
Hexanoic	6	$CH_3(CH_2)_4COOH$
Octanoic	8	$CH_3(CH_2)_6COOH$
Decanoic	10	$CH_3(CH_2)_8COOH$
Dodecanoic (Lauric)	12	$CH_3(CH_2)_{10}COOH$
Myristic	14	$CH_3(CH_2)_{12}COOH$
Palmitic	16	$CH_3(CH_2)_{14}COOH$
Stearic	18	$CH_3(CH_2)_{16}COOH$

2. Unsaturated Fatty Acids

Name	Nbr of Double Bonds	Structure
Oleic	1	cis-9-
Linoleic	2	cis,cis-9,12-
Linolenic	3	all cis-9,12,15-
Arachidonic	4	all cis-5,8,11,14-

3. Comments on Fatty Acids: Cis-9 for oleic acid means that the double bond is between the 9th and 10th carbons (numbering the carbons starting with the carboxyl carbon as 1) and the double bond has the **cis** geometric **configuration**:

$$
\begin{array}{ccc}
H & & H \\
\backslash & & / \\
C & = & C \\
/ & & \backslash \\
CH_3(CH_2)_7 & & (CH_2)_7COOH
\end{array}
$$

The most common fatty acids found in mammals are straight-chain, unsubstituted, monocarboxylic acids and have an even number of carbons, usually 12-20. When a fatty acid contains more than one double bond, it is polyunsaturated. Polyunsaturated fatty acids (linoleic, linolenic, arachidonic) are essential in the diet of mammals.

B. Fatty Acid β-Oxidation

1. Steps in β-Oxidation: After initial **activation** (1) of fatty acid in cytoplasm, the entrance of acyl-CoA into mitochondria requires the **carnitine transport** system. The β-oxidation of acyl-CoA involves 4 reaction steps: **dehydrogenation** (2), **hydration** (3), **dehydrogenation** (4), and **cleavage** (5). This reaction pathway is shown in Figure 7.1.

2. Cofactor Requirement: CoA, carnitine, NAD , FAD and ATP (initially).

3. Stoichiometry: The stoichiometry for the oxidation of palmitate ($C_{16}H_{32}O_2$) can be calculated as shown below:

Palmitate
7 turns of β-oxidation   (7 x 17 ATP)   + 119 ATP
1 acetyl group left   + 12 ATP
1 ATP used for initial activation, but
   cleavage of PP$_1$ costs an extra ~P   -   2 ATP

Net: 129 ATP

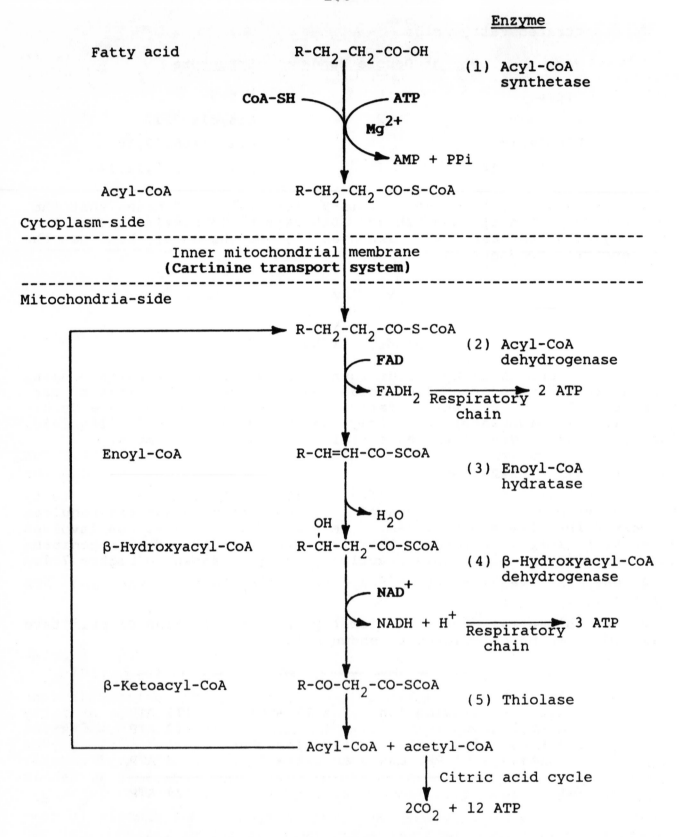

Fig. **7**.1 β-Oxidation of fatty acid in mitochondria.

C. <u>Catabolism of Odd-numbered-carbon Fatty Acids</u>

Oxidation of an odd-numbered-carbon fatty acid yields successive molecules of acetyl-CoA and 1 equivalent of **propionyl-CoA**. The major pathway of propionyl-CoA metabolism is summarized in the following three reactions:

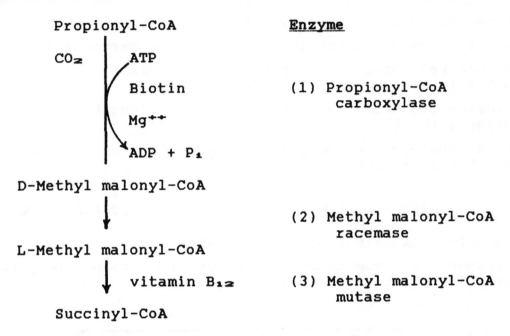

In these enzyme reactions, biotin is a cofactor for propionyl-CoA carboxylase and **vitamin B$_{12}$** is a cofactor for methyl malonyl-CoA mutase. Succinyl-CoA can further enter into the TCA cycle.

D. <u>Biosynthesis of Saturated Fatty Acids</u>.

1. Malonyl-CoA formation: The major *de novo* route is in cytoplasm. The first step of fatty acid synthesis is the formation of **malonyl-CoA** from acetyl-CoA catalyzed by **acetyl-CoA carboxylase.**

$$CH_3-\overset{\overset{\text{O}}{\|}}{C}-S-CoA + ATP + HCO_3^- \xrightarrow{\text{Biotin}} \text{Malonyl-CoA} + ADP + P_i$$

The enzyme contains biotin as the cofactor in the reaction. Biotin-containing enzymes are inhibited by a protein in egg white, avidin.

2. Reaction pathway of fatty acid synthetase (FAS) multienzyme complex (palmitate synthetase): There appear to be two types of fatty acid synthetase systems found in cytoplasm of the cell. In bacteria, plants, and lower forms, the individual enzymes of the system may be separate and the acyl radicals are in combination with acyl carrier protein (ACP). In yeast, mammals and birds, the synthetase system is a **multienzyme complex** that may not be subdivided.

The fatty acid synthetase multienzyme complex of mammals is composed of a dimer with the monomer molecular weight of 267,000. It appears that all of the 7 enzymes of fatty acid synthetase and an ACP reside in one polypeptide chain. The two monomers are aligned in a

cyclic head-to-tail fashion which allows the continuous chain elongation to occur. The thiol group in 4'-phosphopantetheine (Pan) of ACP
is in close proximity to the thiol group of a cysteine residue attached to ß-ketoacyl synthetase of the other monomer. Because both
thiols participate in the synthetase reaction, only the dimer is active. The pathway of fatty acid synthesis is shown in Figure 7.3.

   3. Comments on fatty acid synthesis:

     a. Proceeds for fatty acids of up to 16 carbon atoms.

     b. **Control** step is **acetyl—CoA carboxylase.** The enzyme is in-
ducible and is **activated by citrate.**

     c. Requires **NADPH** when it is produced adequately from the hexose
monophosphate shunt.

     d. Malonyl-CoA carbons always added to carboxyl end of fatty
acid and CO_2 is released.

$$\text{R-COACP} + {}^-\text{OOCCH}_2\text{CO-CoA} \longrightarrow \text{RCOCH}_2\text{COACP} + CO_2$$

E. Fatty Acid Synthesis from Carbohydrate:

   1. Lipogenesis from carbohydrate requires the participation of **mitochondria** in the overall pathway (Figure 7.2).

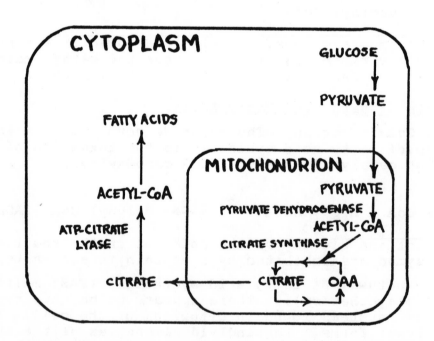

FIGURE 7.2 MITOCHONDRIAL LIPOGENESIS FROM CARBOHYDRATE

   2. The transport of acetyl-CoA from mitochondrion to cytoplasm is
mediated **via citrate** (formation catalyzed by citrate synthetase).

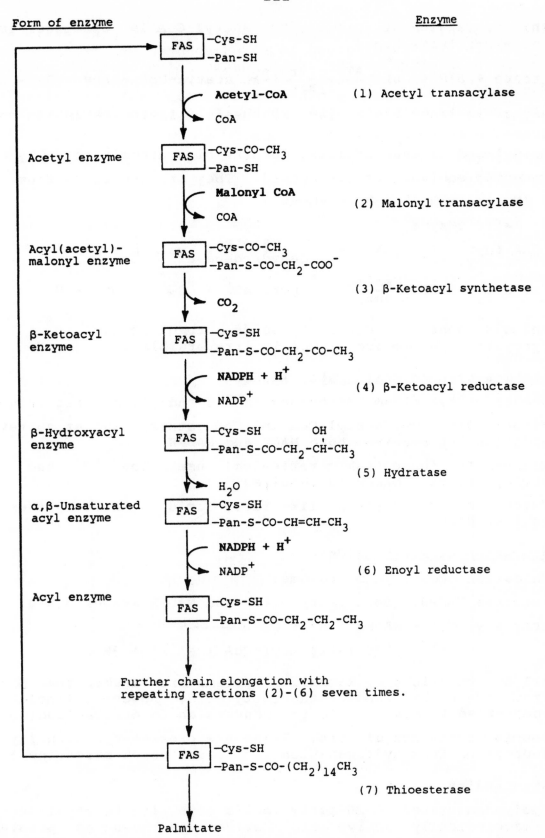

Fig. 7.3 Fatty acid synthetase reactions.

3. The conversion of citrate to acetyl-CoA in cytoplasm is catalyzed by **ATP-citrate lyase**.

$$\text{Citrate + ATP + CoA} \xrightarrow[\text{lyase}]{\text{ATP-citrate}} \text{acetyl-CoA + ADP + Pi + OAA}$$

This enzyme is inducible. Its synthesis is increased in caloric excess.

4. **Acetyl-CoA** is then utilized for the biosynthesis of fatty acids.

5. **Reducing equivalents** are needed. They are generated from:

    a. **Hexose monophosphate shunt**

    b. **Malic enzyme**.

$$\text{OAA (cytosol) + NADH + H}^+ \xrightarrow[\text{dehydrogenase}]{\text{malate}} \text{malate + NAD}^+$$

$$\text{malate + NADP}^+ \xrightarrow[\text{enzyme}]{\text{malic}} \text{pyruvate + CO}_2 \text{ + NADPH + H}^+$$

6. Animals cannot convert fatty acids into glucose. (Pyruvate $\longrightarrow$ acetyl-CoA is irreversible.)

F. Elongation of Long Chain Fatty Acids.

1. Occurs in cytoplasm, mitochondria and endoplasmic reticulum.

2. Mitochondria use acetyl-CoA and CoA esters, similar to reversal of ß-oxidation but requires both NADH and NADPH.

3. Microsomes (endoplasmic reticulum) use malonyl CoA and CoA ester, not ACP-esters. NADPH is required.

4. Cytoplasm uses systems like fatty acid *de novo* synthesis and also requires NADPH.

G. Desaturation of Fatty Acids.

1. Oxidation occurs in microsome.

2. Requires NADPH. Both fatty acids and NADPH are oxidized.

$$\text{Fatty acyl-CoA + NADPH + H}^+ \text{ + O}_2 \longrightarrow$$

$$\text{Unsaturated fatty acyl-CoA + NADP}^+ \text{ + 2H}_2\text{O}$$

3. All desaturations by mammalian systems are **farther than 6 carbon atoms** from the methyl end of the fatty acid. However, linoleic acid can be converted to arachidonic for conversion to prostaglandins.

4. Double bonds are all **cis**. (Trans are present occasionally from plant sources or from hydrogenation in production of margarine.)

H. Prostaglandins

The polyunsaturated C-20 fatty acids give rise to physiologically and pharmacologically active acid derivatives known as **prostanoids** [prostaglandins (PG), prostacyclins (PGI), thromboxanes (TX)], and **leucotrienes**(LT).

The major classes of prostaglandins (as they are known collectively) are designated with a numerical **subscript** which denotes the **number of double bonds** in the molecule. Because of the cyclization reaction, prostanoids have 2 fewer double bonds than the parent fatty acid, while leucotrienes have the same number as the parent acid. The 3 groups of eicosanoids, their biosynthetic origins, and some examples are shown below.

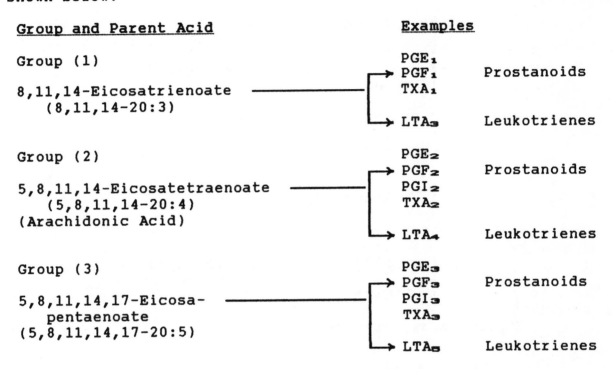

<u>Group and Parent Acid</u>		<u>Examples</u>	
Group (1)		PGE_1 PGF_1 TXA_1	Prostanoids
8,11,14-Eicosatrienoate (8,11,14-20:3)		LTA_3	Leukotrienes
Group (2)		PGE_2 PGF_2 PGI_2 TXA_2	Prostanoids
5,8,11,14-Eicosatetraenoate (5,8,11,14-20:4) (Arachidonic Acid)		LTA_4	Leukotrienes
Group (3)		PGE_3 PGF_3 PGI_3 TXA_3	Prostanoids
5,8,11,14,17-Eicosapentaenoate (5,8,11,14,17-20:5)		LTA_5	Leukotrienes

Prostanoid synthesis (cyclooxygenase pathway) involves the consumption of 2 molecules of O_2 catalyzed by **prostaglandin endoperoxide** which possesses 2 enzyme activities, cyclooxygenase and peroxidase. **Aspirin** was found to **inhibit** the **cyclooxygenase** reaction. Thromboxanes are synthesized in platelets and upon release cause **vasoconstriction** and **platelet aggregation**. Prostacyclins (PGI_2) are produced by blood vessel walls and are potent **inhibitors of platelet aggregation**.

Leucotrienes are a newly-discovered family of conjugated trienes formed from eicosanoic acids by the lipoxygenase pathway. The slow-reacting substrate of anaphylaxis is a mixture of LTC_4, LTD_4, LTE_4. These **leucotrienes** also cause vascular permeability and are important in **inflammatory** and **immediate hypersensitivity** reactions.

V. KETONE BODIES

A. <u>Synthesis of Ketone Bodies</u>.

Ketogenesis is a hepatic process (Figure 7.4). In liver mitochondria, fatty acids are oxidized to **acetyl CoA**, a portion of which is oxidized via the TCA-cycle to CO2 and a portion is converted to ketone bodies via **3-hydroxy-3-methylglutaryl–CoA** (HMG-CoA).

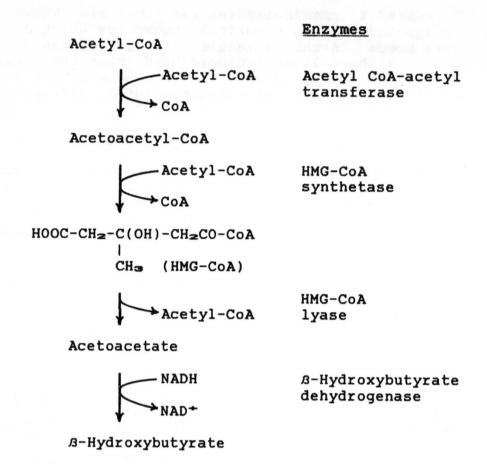

Enzymes

Acetyl-CoA

Acetyl-CoA → CoA   Acetyl CoA-acetyl transferase

Acetoacetyl-CoA

Acetyl-CoA → CoA   HMG-CoA synthetase

$HOOC-CH_2-C(OH)-CH_2CO-CoA$
|
CH_3 (HMG-CoA)

→ Acetyl-CoA   HMG-CoA lyase

Acetoacetate

NADH → NAD^+ ß-Hydroxybutyrate dehydrogenase

ß-Hydroxybutyrate

Figure 7.4. Hepatic Ketogenesis.

B. Metabolism and Regulation

1. Utilization: Ketone bodies are utilized primarily by **muscle**.

2. Starvation: The liver glycogen in the rat is depleted by about 70% after a 12-hour fast and is almost completely depleted after 24 hrs of starvation. Acetyl-CoA derived from fatty acid ß-oxidation is channeled into ketogenesis. The carbohydrate and fatty acid synthetic pathways are inhibited. Normally, adult brain depends entirely on glucose for its energy. Under conditions of long term **starvation** (more than 3 days), the **brain** derives a considerable part of its energy from these acids.

3. Insulin deficiency: From the point of view of lipid metabolism, diabetes and starvation resemble each other. Fatty acids mobilized from adipose tissue raise the level of acetyl-CoA in the liver and promote **ketogenesis.**

4. Consequences of ketosis: Ketone bodies are excreted in large part as the sodium salts; depletion from the body fluids of Na+ and other cations leads to **acidosis** (ketoacidosis).

VI. TRIACYLGLYCEROLS

A. <u>Synthesis</u> *de novo*, in liver and adipose tissue (Figure 7.5).

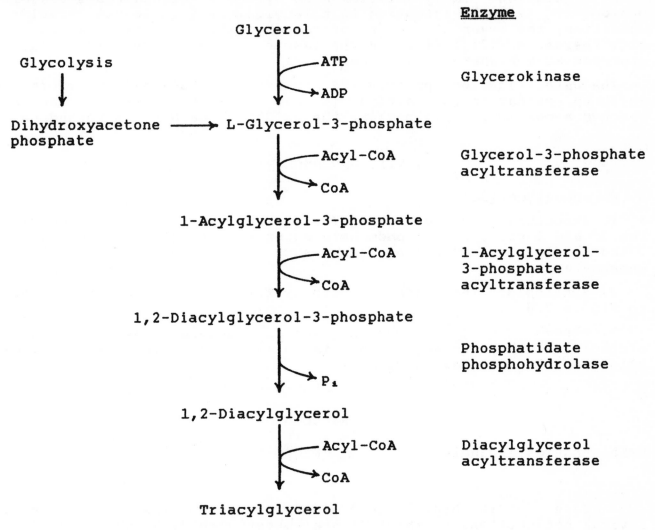

<u>**Figure 7.5**</u>. Formation of Triacylglycerol.

B. <u>Utilization</u>

  1. Digestion: Triacylglycerols in the diet are hydrolyzed in the intestine, resynthesized by the intestinal cells, then transferred to the lymph as **chylomicrons**.   Intestinal protein synthesis is required for the secretion of chylomicrons.

  2. Transport: from liver as **very low density lipoproteins** (VLDL).

  3. **Lipoprotein lipase:** bound to capillary endothelium. Peripheral tissues require it in order to utilize the triacylglycerol from triacylglycerol-containing lipoproteins.

4. **Hormone-sensitive lipase** (HSL): hydrolyzes triacylglycerols to fatty acids and glycerol during fat mobilization in adipose tissue.

5. Regulation: fatty acid taken up by the liver undergoes (1) **oxidation**, or (2) **esterification** to triacylglycerols. In the normal fed condition, the liver has plenty of carbohydrate to be oxidized for energy therefore (2) > (1). In the fasting state, fatty acid oxidation proceeds at a higher rate, therefore (1) > (2).

The esterification pathway (2) is never saturated, therefore the shift to oxidation in fasting is not because esterification has reached a maximum, it is because the oxidative pathway is turned on.

VII. PHOSPHOLIPIDS

A. Phosphoglycerides

1. Structure: 2 long-chain **fatty acids** esterified to glycerol in the 1 and 2 positions, a **phosphate** esterified to glycerol in position 3 and a nitrogen-containing **base**, such as choline, esterified to the phosphate.

2. Synthesis: The synthetic pathway for phosphoglycerides is shown in Figure 7.6.

3. Plasmalogen: These phosphoglycerols contain an **α,ß-unsaturated alcohol** in **ether linkage**. Thus, the plasmalogens containing ethanolamine have the general structure:

$$H_2C-O-CH=CHR_1$$
$$|$$
$$R_2COOCH \quad O^-$$
$$| \quad |$$
L-Phosphatidalethanolamine: $\quad H_2C-O-P-OCH_2CH_2NH_2$
$$\overset{"}{O}$$

4. Sites for the Hydrolytic Activity of Phospholipases: There are several **phospholipases** which cleave phospholipids at specific places in the molecule. They are useful in investigations requiring highly specific cleavage of phospholipids and phosphate esters.

B. Sphingomyelin

1. Structure:

a. **Ceramide** (N-Acyl-Sphingosine):

$$OH$$
$$|$$
$$CH_3(CH_2)_{12}-CH=CH-CH-CH-CH_2OH$$
$$|$$
$$HN-C-R$$
$$\overset{"}{O}$$

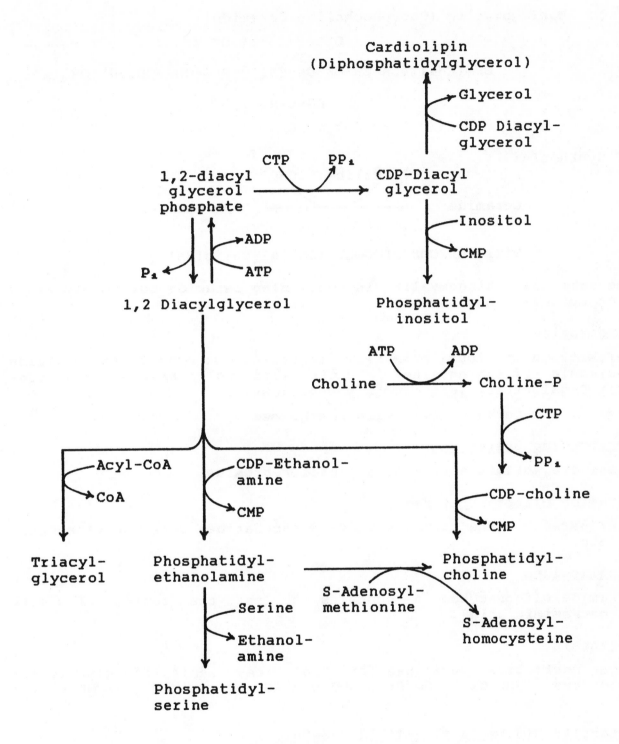

Figure 7.6. Synthetic Pathway for Phosphoglycerides.

b. **Sphingomyelin** (Phosphocholine Ceramide):

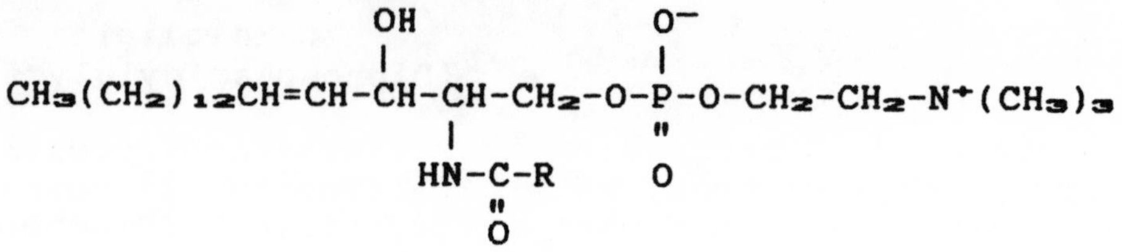
OH
O⁻
CH3(CH2)12CH=CH-CH-CH-CH2-O-P-O-CH2-CH2-N+(CH3)3
HN-C-R
O
O

2. Synthesis:

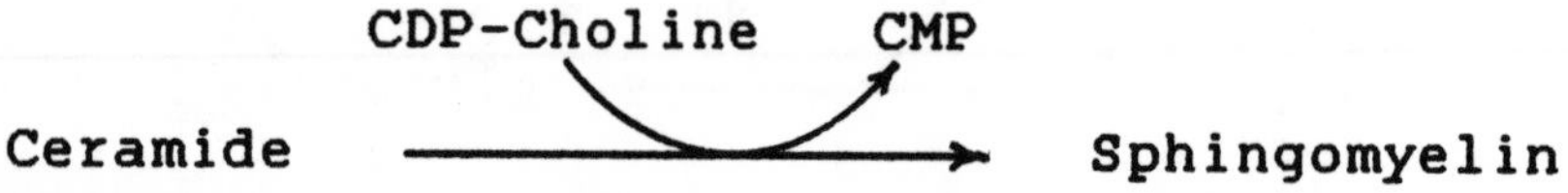
CDP-Choline
CMP
Ceramide
Sphingomyelin

VIII. GLYCOSPHINGOLIPIDS (glycolipids).

The same as sphingomyelin in containing ceramide but no glycerol. They do not contain phosphate.

A. Cerebroside

Cerebroside is found primarily in brain. On hydrolysis it yields one molecule of sphingosine, a fatty acid and a hexose (most frequently D-galactose, less commonly D-glucose).

Cerebroside: **Ceramide-hexose**

B. Cerebroside Sulfatides

These are **sulfate** esters of **galactocerebroside**.

C. Ceramide Oligosaccharides

Designated by the terms ceramide disaccharide, ceramide trisaccharide, etc.

D. Gangliosides

Ceramide oligosaccharide contains at least one residue of **sialic acid (neuraminic acid)**.

E. Synthesis

Sugar nucleotides utilized for the biosynthesis of glycosphingolipids are: UDP-Glc, UDP-Gal, UDP-GlcNAC, UDP-GalNAC and CMP-sialic acid.

F. Metabolic Blocks in Glycolipid Catabolism:

Several **lysosomal disorders** exist, in which various glycolipids accumulate ("lipid storage diseases"). See Chapter 11.

IX. LIPOPROTEINS

A. Lipid Transport in Blood

Lipids are insoluble in water and so must be transported as lipoproteins. Lipoproteins are classified according to increasing density: chylomicrons, very low density lipoproteins, low density lipoproteins and high density lipoproteins. Nine apolipoproteins (Apo A-I, A-II, A-IV, B, C-I, C-II, C-III, D and E) have been isolated and characterized.

B. Chylomicrons

Chylomicrons are the **largest** of the lipoprotein particles which transport dietary triacylglycerols, cholesterol and other **lipids** from the **intestine** to adipose tissue and the liver. The triacylglycerols in chylomicrons are hydrolyzed within a few minutes by **lipoprotein lipase.** The cholesterol-rich residue particles, known as **remnants**, are taken up by the liver. They are especially abundant in blood after a carbohydrate-rich meal and conditions of caloric excess.

C. Very Low Density Lipoproteins (VLDL)

VLDL are primarily synthesized by the **liver.** VLDL are also degraded via lipoprotein lipase, resulting in the production of **low density lipoproteins.**

D. Low Density Lipoproteins (LDL)

LDL are rich in **cholesterol ester.** These lipoproteins are **atherogenic.** The removal of LDL is via a specific **receptor-mediated** process. Some is also removed via scavenger processes.

Cells outside the liver and intestine obtain cholesterol from the plasma rather than by synthesizing it *de novo.* The steps in the uptake of LDL are: 1) LDL **binds** to a plasma membrane region called **coated pits.** 2) Receptor-LDL complex is **internalized by endocytosis.** 3) These vesicles are subsequently fused with **lysosomes** and degraded. 4) The cholesterol content of cells that have an active LDL pathway is regulated in two ways. First, the released **cholesterol suppresses** the formation of 3-hydroxyl-3-methylglutaryl-CoA reductase (**HMG-CoA reductase**), which thereby inhibits the *de novo* synthesis of cholesterol. Second, LDL receptor is regulated with a feedback mechanism. When cholesterol is abundant inside the cell, **new LDL receptors are not synthesized** and the uptake of additional cholesterol from plasma LDL is blocked.

The **genetic defect** in most cases of familial hypercholesterolemia is due to an absence or deficiency of the normal **receptor for LDL.**

E. High Density Lipoproteins (HDL)

HDL are produced by the **liver.** They are involved in the transport of cholesterol from the periphery to the liver. They appear to be **antiatherogenic.**

F. Lipoprotein Lipase (LPL)

Lipoprotein triacylglycerols in chylomicrons and VLDL cannot be taken up intact by tissues but must first undergo **hydrolysis** by LPL, an enzyme situated on the capillary endothelium of extrahepatic tissues. The enzyme requires apolipoprotein C-II as cofactor.

G. Hormone Sensitive Lipase (HSL)

The fat reserves of mammals are stored in **adipose tissue**. The triacylglycerols must first be **hydrolyzed** to **free fatty acid** before they can be transported from the adipose tissue to other tissues. The hydrolysis of triacylglycerols within adipose tissue is catalyzed by hormone sensitive lipase. The released free fatty acid is transported in blood by the plasma protein, **albumin**.

The lipolysis of adipose tissue fat (fat mobilization) is stimulated by **glucagon** and **epinephrine**. Lipolysis is inhibited by insulin, which stimulates glucose uptake for lipogenesis, decreasing cAMP in fat cells.

H. Hypo- and Hyperlipoproteinemias

A number of heritable disorders of lipoprotein metabolism occur. (See Chapter 11).

I. Hypolipidemic Drugs.

Cholestyramine binds with bile salts and prevents their reabsorption. **Clofibrate** acts in part by decreasing the synthesis of triacylglycerols and their secretion into the bloodstream. Other drugs that are considered to increase the fecal excretion of cholesterol and bile acids include **dextrothyroxine** (Choloxin), **neomycin**, and **probucol**.

X. REVIEW QUESTIONS ON LIPIDS

DIRECTIONS: Each of the questions or incomplete statements below is followed by five suggested answers or completions. Select the one that is BEST in each case and fill in the corresponding space on the answer sheet.

1. Glucose catabolism stimulates fatty acid synthesis primarily because it

A. increases acetyl-CoA
B. increases NADH
C. increases NADPH
D. increases ATP
E. increases glycogen

2. Which of the following is the most complex CoA ester on the major route to ketone bodies in liver?

A. acetyl-CoA
B. malonyl-CoA
C. acetoacetyl-CoA
D. mevalonyl-CoA
E. hydroxymethyl glutaryl-CoA

3. An odd-numbered-carbon-atom fatty acid will enter the citric acid cycle in part as

A. citrate
B. isocitrate
C. α-ketoglutarate
D. succinate
E. malate

4. In the reaction pathway of pyruvate to acetyl-CoA, which of the following enzyme reactions occurs in cytoplasm?

A. pyruvate dehydrogenase
B. citrate synthetase
C. propionyl-CoA carboxylase
D. succinate dehydrogenase
E. ATP-citrate lyase

5. The role of lipoprotein lipase is for:

A. digestion of dietary lipoproteins in the intestinal lumen
B. mobilization of dietary fat
C. intracellular lipolysis of lipoproteins
D. hydrolysis of triacylglycerols from plasma lipoprotein for transport of the released fatty acids into tissues.
E. none of the above.

6. The mitochondrial membranes are permeable to

A. fatty acyl ACP
B. fatty acyl-CoA
C. malonyl-CoA
D. acetyl-CoA
E. none of the above

7. Increased acetoacetate might be expected in urine when liver

A. glycogen is normal
B. glycogen is depleted
C. NADPH concentration is high
D. acetyl CoA concentration is low
E. none of the above

8. The diet must provide

A. lecithin
B. sphingomyelin
C. cerebrosides
D. linoleic acid
E. phosphatidic acid

9. Carbon atoms from fatty acids would be expected to be in highest amounts in

A. glucose
B. alanine
C. aspartic acid
D. glycerol
E. all would contain equal amounts

10. All glycerol-containing lipids are synthesized from

A. triglyceride
B. cephalin
C. phosphatidic acid
D. diglyceride
E. monoglyceride

11. The conversion of a diglyceride to lecithin requires

A. UDP-glucose
B. CDP-choline
C. ACP-fatty acid ester
D. malonyl-CoA
E. none of the above

12. The role of hormone sensitive lipase is to

A. hydrolyze the ester bonds in hormones
B. hydrolyze dietary fat and the enzyme is stimulated by epinephrine
C. mobilize fat from adipose tissue
D. hydrolyze triacylglycerols in liver
E. hydrolyze triacylglycerols in brain

13. The complete hydrolysis of cardiolipin results in the release of

A. 0 mole of fatty acid
B. 1 mole of fatty acid
C. 2 moles of fatty acid
D. 3 moles of fatty acid
E. 4 moles of fatty acid

14. The major energy source for the brain is normally

A. blood glucose
B. blood amino acid
C. blood ketone bodies
D. blood fatty acids
E. blood lactic acid

15. Chylomicron triglycerides

A. are intestinal lumen triglycerides.
B. are newly synthesized by the intestinal cell.
C. are transported from the liver
D. none of the above
E. all of the above

16. Acetyl-CoA carboxylase

A. is activated by citrate.
B. is the rate limiting step in fatty acid synthesis.
C. contains biotin
D. all of the above
E. none of the above

17. The fate of $^{14}CO_2$ incorporated into the carboxyl of malonyl-CoA during the condensation step as carried out by palmitate synthetase is

A. incorporation into the fatty acid.
B. incorporation into acetyl-CoA being synthesized.
C. elimination as $^{14}CO_2$.
D. none of the above
E. all of the above

18. Ketosis is the consequence of increased blood levels of

A. acetoacetate.
B. acetyl-CoA
C. β-hydroxy-β-methyl-glutarate.
D. none of the above.
E. all of the above.

19. Fatty acids in transport from adipose tissue to energy-utilizing tissues like muscle occur in the blood in the form of

A. chylomicrons.
B. free fatty acids bound to albumin.
C. VLDL
D. HDL
E. all of the above

20. Accumulation of lipids in lipidoses, such as Tay-Sachs disease, is primarily a result of

A. increased synthesis of glycolipids.
B. decreased breakdown of glycolipids.
C. increased synthesis and decreased turnover.
D. none of the above
E. all of the above

21. Gangliosides contain

A. neuraminic acid
B. galactose
C. fatty acid
D. sphingosine
E. all of the above

22. How many grams of stearate must be metabolized to give an amount of energy equivalent to 50 g of glycogen?

A. 5
B. 20
C. 40
D. 80
E. 94

23. Acetyl-CoA carboxylase and other biotin-containing enzymes are inhibited by

A. citrate
B. carnitine
C. avidin
D. lactalbumin
E. cyanide

DIRECTIONS: For each of the questions or incomplete statements below, ONE or MORE of the answers or completions is correct. On the answer sheet fill in space

A   if only 1, 2, and 3 are correct
B   if only 1 and 3 are correct
C   if only 2 and 4 are correct
D   if only 4 is correct
E   if all are correct

FILL IN ONLY ONE SPACE ON YOUR ANSWER SHEET FOR EACH QUESTION

Directions Summarized				
(A) 1,2,3 only	(B) 1,3 only	(C) 2,4 only	(D) 4 only	(E) All are correct

24. The cofactors common to both β-oxidation and fatty acid synthesis include

1. FAD
2. NAD
3. NADP
4. CoA

25. The ketone bodies are mainly produced from

1. phospholipids in liver
2. triglycerides of fat cells
3. cholesterol
4. plasma free fatty acids

26. In the fasting state, the oxidation of fatty acids or ketone bodies leads to a slower rate of glycolysis in muscle because

1. acetyl-CoA inhibits pyruvate dehydrogenase
2. the hexokinase reaction becomes rate-limiting owing to the increased ATP/ADP ratio.
3. the low insulin level diminishes the uptake of glucose by muscle
4. the increased NADH/NAD+ ratio decreases the activity of glyceraldehyde-3-phosphate dehydrogenase

27. The de novo biosynthesis of triacylglycerols occurs mainly in

1. liver
2. brain
3. adipose tissue
4. muscle

28. The cofactors common to both β-oxidation of odd-numbered fatty acids and fatty acid synthesis include

1. vitamin B_{12}
2. coenzyme A
3. NADP+
4. biotin

29. Fat mobilization in fat cells is enhanced by

1. decrease in cAMP
2. increase in cAMP
3. increase in insulin
4. increase in epinephrine

30. An increase in citrate concentration in cytoplasm would cause:

1. slowing of glycolysis by inhibiting the activity of phosphofructokinase.
2. enhancement of glycolysis by stimulating phosphofructokinase.
3. increase in fatty acid synthesis.
4. none of the above

31. Endocytosis of LDL-receptor vesicles and subsequent degradation by lysosome leads to

1. increased activity of hydroxymethylglutaryl-CoA reductase.
2. decreased activity of hydroxymethylglutaryl-CoA reductase.
3. increased synthesis of LDL-receptor.
4. decreased synthesis of LDL-receptor.

FILL IN ONLY ONE SPACE ON YOUR ANSWER SHEET FOR EACH QUESTION

Directions Summarized				
(A) 1,2,3 only	(B) 1,3 only	(C) 2,4 only	(D) 4 only	(E) All are correct

32. Products of the oxidation of fatty acids containing an odd number of carbons include

1. propionyl-CoA
2. malonyl-CoA
3. acetyl-CoA
4. none of the above

33. Essential fatty acids and precursors of prostaglandins include:

1. linoleic acid
2. lignoceric acid
3. linolenic acid
4. oleic acid

34. Compounds formed during the catabolism of palmitic acid and which are intermediates for the formation of ketone bodies include

1. palmitoyl carnitine
2. acetyl-CoA
3. acetoacetyl-CoA
4. ß-hydroxy ß-methylglutaryl-CoA

35. In addition to ATP activation, the steps involved in the palmitate synthetase system include

1. dehydration of a double bond
2. decarboxylation
3. reduction of a carbonyl
4. condensation

36. Transport of the acyl moiety of acetyl-CoA, which is produced in mitochondria, to the cytosol for palmitate synthesis requires

1. carnitine acyl transferase
2. acetyl-CoA carboxylase
3. CoA-hydrolase
4. citrate cleavage enzyme

37. Substances requiring bile salt for transport into intestinal cells include:

1. vitamin A
2. vitamin C
3. vitamin D
4. glucose

38. Cholestyramine exerts its hypocholesterolemic effect by interaction with:

1. cholesterol
2. triacylglycerols
3. lipoproteins
4. bile salts

DIRECTIONS: Each group of questions below consists of five lettered headings followed by a list of numbered words or statements. For each numbered word or statement, select the one lettered heading that is most closely associated with it and fill in the corresponding space on the answer sheet. Each heading may be selected once, more than once, or not at all.

Questions 39-41:

A. ATP
B. CTP
C. GTP
D. ADP
E. AMP

39. Required for activation of fatty acid.

40. Required for activation of diglyceride.

41. Required for activation of choline.

Questions 42-43.

A. phosphate
B. choline
C. fatty acid
D. sphingosine
E. glycerol

42. Hydrolysis of lecithin would yield all of the above EXCEPT which?

43. Hydrolysis of sphingomyelin would yield all of the above EXCEPT which?

XI. ANSWERS TO QUESTIONS ON LIPIDS

1. C		23. C	
2. E		24. D	
3. D		25. C	
4. E		26. B	
5. D		27. B	
6. E		28. C	
7. B		29. C	
8. D		30. B	
9. C		31. C	
10. C		32. B	
11. B		33. B	
12. C		34. E	
13. E		35. E	
14. A		36. D	
15. B		37. B	
16. D		38. D	
17. C		39. A	
18. A		40. B	
19. B		41. B	
20. B		42. D	
21. E		43. E	
22. B			

8. STEROIDS

Thomas Briggs

I. CHOLESTEROL: STRUCTURE AND CHEMISTRY

Carbon skeleton: four rings plus 8-carbon side-chain = 27 carbons:

HO

Substitution: ... (dotted line) = α (alpha, away from observer).
—— (solid line) = ß (beta, toward, or by observer).

The **3ß-hydroxyl** group, though hydrophilic, cannot overcome the non-polarity of the rest of the molecule, hence cholesterol is **insoluble in water.**

The Δ^{5-6} double bond can be reduced two ways:
   5α-H: 5α-cholestan-3ß-ol; new hydrogen _trans_ to the angular methyl group (C-19); rings A/B _trans_.
   5ß-H: 5ß-cholestan-3ß-ol; new hydrogen _cis_ to C-19; rings A/B _cis_. This is mainly a bacterial product (coprosterol in older usage, from Gr. _copros_, meaning feces, where it is found).

II. OCCURRENCE AND FUNCTION

The name means **"bile-solid-alcohol."** It is found in bile, (a frequent component of gallstones, where it was first discovered), and is a crystalline solid and an alcohol.

Cholesterol occurs in **all cells** and tissues of higher organisms, but is especially abundant in nervous tissue and egg yolk. Related sterols are found in plants and higher microorganisms; these are poorly absorbed from the digestive tract. For storage, sterols are often **esterified** with unsaturated fatty acids. The total amount in a human averages about 180 grams. A typical value for blood cholesterol in developed countries is 180 mg/100 ml. Note that on a weight basis this is twice as high as the level of blood glucose.

Cholesterol has a universal function as a **component of membranes**, probably to increase their fluidity. Also it is a **precursor** of many important biological substances (see **bile acids, hormones**). In human disease, cholesterol's insolubility makes deposits troublesome, especially in **atherosclerotic plaques** and **gallstones**.

In blood, cholesterol is carried as a complex with **lipoproteins**, partly esterified with a fatty acid, and partly with the 3-OH free. (See Chapter 7, page 119). The cholesterol in blood can exchange readily with liver cholesterol, but some cholesterol pools, especially that in brain, exchange very slowly.

III. BIOSYNTHESIS OF CHOLESTEROL

A. Acetate to Squalene

   1. Formation of **ß-hydroxy-ß-methyl-glutaryl CoA** (HMG CoA)

      a. 2 acetyl CoA $\rightleftharpoons$ acetoacetyl CoA + CoA (as in ketogenesis)

      b. acetoacetyl CoA + acetyl CoA $\rightleftharpoons$ ß-hydroxy-ß-methyl-glutaryl CoA, or HMG CoA

   2. Reduction of HMG CoA to **Mevalonic Acid**

$$\text{HMG CoA} + 2\text{NADPH} + 2\text{H}^+ \xrightarrow[\text{Reductase}]{\text{HMG CoA}} \text{MVA} + 2\text{NADP}^+ + \text{HSCoA}$$

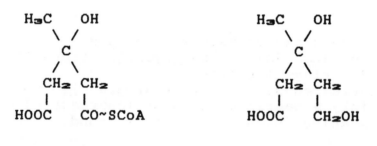

HMG CoA   Mevalonic Acid (MVA)

This reaction is the first committed, largely irreversible step in cholesterol biosynthesis, and is an **important control point**. The enzyme is rate-limiting for the entire pathway, and is inhibited by the eventual end-product, cholesterol.

   3. MVA $\xrightarrow[\text{3 times}]{\text{ATP, Mg}^{++}}$ $\text{H}_2\text{C}=\text{C-CH}_2\text{-CH}_2\text{OPP}$ **Isopentenyl**
                                        | **Pyrophosphate**
                                       CH_3 (CO_2 is lost)

Isopentenyl pyrophosphate is the "active isoprene unit": the precursor of all isoprenoid compounds such as terpenes; vitamins A, D, E, K; sterols; rubber.

4. Isopentenyl-PP can isomerize to **dimethylallyl-PP,** which conden-ses with another isopentenyl-PP to form **geranyl-PP,** a monoterpene (C-10):

$$H_2C=C-CH_2-CH_2OPP \rightleftharpoons H_3C-C=CH-CH_2OPP \longrightarrow H_3C-C=CH-CH_2CH_2C=CH-CH_2OPP$$

$$\qquad\; CH_3 \qquad\qquad\qquad\quad CH_3 \qquad\qquad\qquad\quad CH_3 \qquad\quad CH_3$$

5. Geranyl-PP condenses with another i-pentenyl-PP to form **farnesyl-PP,** a sesquiterpene (C-15).

6. Two farnesyl-PP condense head-to head (NADPH required) to form **squalene,** a triterpene (C-30).

B. <u>Squalene to Cholesterol</u> (Figure 8.1)

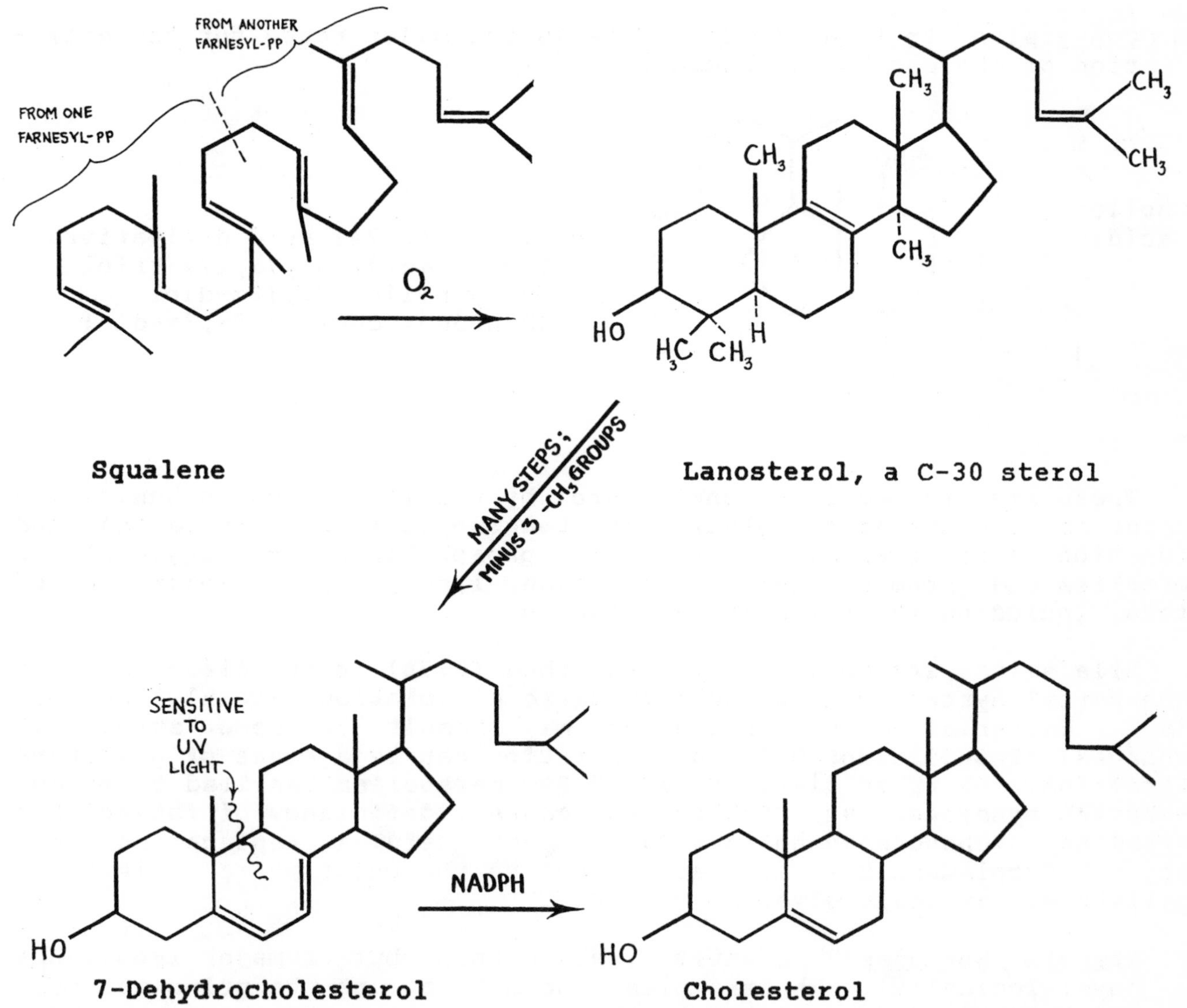

Figure 8.1. Conversion of Squalene to Cholesterol

Ring B of 7-dehydrocholesterol can be split by UV light (at arrow, Figure 8.1) to form vitamin D₃. The presence of 7-dehydrocholesterol in human skin explains how sunlight leads to the formation of vitamin D. In the body, vitamin D (cholecalciferol) is transported to the liver and hydroxylated at C-25, then to the kidney and hydroxylated at 1α, to form **1α,25-dihydroxycholecalciferol**, or **calcitriol**. <u>This is the biologically active form.</u> Vitamin D can be regarded as a hormone because it acts on target cells by the same mechanism as the steroid hormones. The active form promotes calcium absorption in the intestine by stimulating the synthesis of a **calcium binding protein**.

IV. METABOLISM OF CHOLESTEROL

A. <u>Secretion</u>: Some is simply **secreted as is** into intestine.

B. <u>Conversion to Bile Acids</u>: This is the major route for the excretion of cholesterol in humans.

Cholic acid:

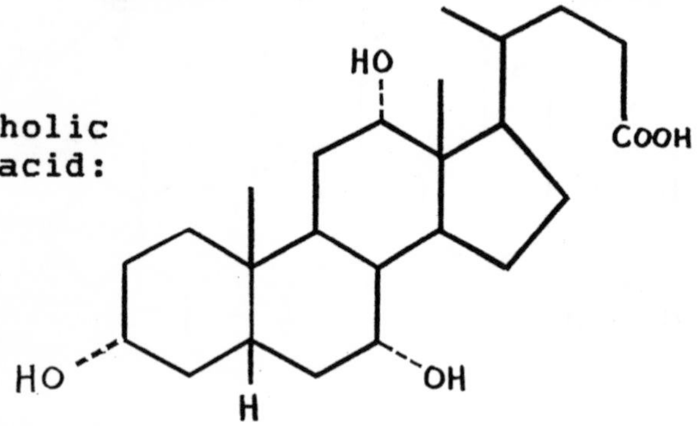

Cholanoic (C-24) acid derivatives:
Cholic acid: 3α,7α,12α-triol
Deoxycholic: 3α,12α-diol
Chenodeoxycholic: 3α,7α-diol

These are formed from cholesterol by the liver, and in human bile occur as conjugates of **glycine** or **taurine** (called bile salts) and function as detergents. Together with phospholipid, they aggregate as **micelles** and promote the emulsification, lipolysis, and absorption of fats, including the fat-soluble vitamins.

Bile salts are extensively reabsorbed (>95%) in the ileum, and via the portal system, undergo **enterohepatic circulation** several times per day. Bacterial action in the gut may result in some structural changes. Insufficient bile salt secretion can be a cause of gallstone formation. Further disturbance of BS metabolism can lead to malabsorption syndromes and, in extreme cases, deficiency of fat-soluble vitamins. Chenodeoxycholic acid has been useful as replacement therapy to supplement the bile acid pool to the point where cholesterol gallstones may redissolve.

C. <u>Steroid Hormones</u>: Quantitatively minor, but of major importance physiologically. The examples shown below are the principal members of each type secreted in the human.

1. Major types
a. adrenal steroids (corticosteroids)
i. glucocorticoids   ii. mineralocorticoids

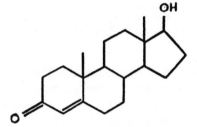

Cortisol   Aldosterone

b. Gonadal steroids
i. progestational   ii. androgens   iii. estrogens

progesterone   testosterone   estradiol

2. Formation from Cholesterol

a. All steroid-producing tissues cleave the side-chain of cholesterol between carbons 20 and 22 to form **pregnenolone**. This is the rate-limiting step. In most cases, pregnenolone is then converted to **progesterone**.

b. Adrenal cortex: hydroxylations occur in sequence at positions **17, 21, 11, 18**. Except: the *zona fasciculata*, which makes cortisol, has no 18-hydroxylase, and the *zona glomerulosa*, which makes aldosterone, has no 17-hydroxylase. Some conversion to androgens (especially **dehydroepiandrosterone**, or **DHEA**) and estrogens may also occur. In a patient with Cushing's syndrome, these androgens may cause signs of virilization such as hirsutism.

c. Gonads: The *corpus luteum* stops at progesterone. The testis, after 17-hydroxylation, cleaves the remaining side-chain to form **19-carbon** steroids (androstenedione, then testosterone). The ovary also does this, in fact makes testosterone, but then aromatizes ring A (resulting in the loss of carbon 19) to form **18-carbon** steroids (estradiol, estrone).

3. Metabolism: reductions occur involving ring A, and the products are excreted in bile and/or urine as conjugates of **sulfate** or **glucuronic acid**. Androgens are oxidized at C-17 to ketones of the class known as **17-ketosteroids**. Urinary 17-KS are indicative of androgen metabolism of <u>both</u> testicular and adrenal origin.

V. ACTION OF STEROID HORMONES

A. <u>Mechanism</u>

In contrast to most peptide hormones, which interact with a receptor on the plasma membrane without penetrating the cell, steroid hormones **enter target cells** and are bound by a **receptor** in the **cytoplasm**. The hormone-receptor complex then **diffuses to the nucleus** and binds to chromatin of selected genes, inducing the **synthesis** (or repression) of particular **proteins**.

B. <u>Function</u>

1. Adrenal hormones

a. **Glucocorticoids** promote gluconeogenesis by inducing key enzymes such as pyruvate carboxylase. Proteins, especially those of muscle, are broken down to amino acids, which the liver converts into glucose which is partly released to the circulation and partly stored as liver glycogen.

Cortisol (also known as hydrocortisone) and synthetic steroids such as prednisone and prednisolone also have anti-inflammatory and anti-immune effects. When used in high doses over a long time they cause, besides muscle wasting, lipolysis on the extremities but accumulation of fat on the face and trunk ("Cushingoid" features).

b. **Mineralocorticoids** promote retention of Na^+, along with H_2O, by the kidney, and excretion of H^+ and K^+. A high level of aldosterone causes hypertension; deficiency, excessive loss of salt.

2. Gonadal Hormones

a. **testosterone**: promotes development of male secondary sex characteristics, and also has anabolic activities. With FSH, it promotes spermatogenesis by seminiferous tubules.

b. **estradiol**: promotes development of female secondary sex characteristics, proliferative phase of endometrium, and, with FSH, development of ovarian follicle and finally, ovulation.

c. **progesterone**: supports secretory phase of endometrium, luteal phase of ovary, and inhibits further ovulation.

C. <u>Regulation</u>

1. Adrenal hormones

a. The hypothalamus puts out **Corticotropin Releasing Hormone** (CRH) which stimulates the pituitary to release **Adrenocorticotrophic Hormone (ACTH)**, which then signals the adrenal cortex to secrete

cortisol. Negative feedback by cortisol occurs on both pituitary and hypothalamus. In extreme cases, such as when steroid drugs are used in high doses for a long time, adrenal **atrophy** may occur due to prolonged inhibition of ACTH production. Conversely, lack of cortisol production, as in an enzymatic defect in the adrenal, may lead to adrenal **hyperplasia** due to excessive ACTH production in an attempt by the pituitary to compensate for the lack of steroid hormone.

b. Production of **aldosterone** is regulated largely by the renin-angiotensin system. In response to a perceived drop in perfusion pressure, the kidney produces **renin**, which converts **angiotensinogen**, a peptide made by the liver, to angiotensin I. Then this is converted by an enzyme in lung tissue to **angiotensin II**, which induces production of aldosterone in the *zona glomerulosa* of the adrenal. Potassium also induces formation of aldosterone. Thus an excess of renin or aldosterone causes hypertension; deficiency results in salt loss.

2. Gonadal hormones: Trophic hormones are the same in both sexes. The hypothalamic **Gonadotropin Releasing Hormone (GnRH)**, which is the same as **Luteinizing Hormone Releasing Hormone (LHRH)**, stimulates the pituitary to release both **Follicle Stimulating Hormone (FSH)**, (formerly known as ICSH) and **Luteinizing Hormone (LH)**.

In the male, FSH induces spermatogenesis in seminiferous tubules; feedback is by a glycoprotein, **inhibin**. LH stimulates production of **testosterone** by Leydig cells. Feedback is on both pituitary and hypothalamus.

In the female, FSH induces follicular development and production of **estradiol**. At mid-cycle, there is a positive feedback effect by estrogen, causing a surge of both FSH and LH production by the pituitary. Luteinization ensues, with production of **progesterone** by the *corpus luteum*.

VI. CHOLESTEROL LEVELS

Since a high level of cholesterol in blood, especially that contained in LDL, is a risk factor in atherosclerotic heart disease, much attention is being given to factors that lower cholesterol levels.

A. Cholesterol-Lowering Agents

1. **Diet**: Cholesterol itself can be avoided by substituting vegetable products for meat and dairy products. In addition, consuming more polyunsaturated fats and fewer saturated fats has a cholesterol-lowering effect.

2. **Exercise**: This activity seems to increase the HDL/LDL ratio, which correlates negatively with atherosclerosis.

3. **Drugs**: Some that have been helpful are: **clofibrate** and **nicotinic acid**, which lower cholesterol levels by poorly-understood mechanisms; **compactin**, which inhibits HMG CoA reductase, the controlling step in cholesterol synthesis; and **cholestyramine**, a resin which, taken orally, binds bile acids and promotes their excretion rather than en-

terohepatic circulation, causing the liver to replace them by new synthesis from cholesterol, some of which comes from the blood.

B. Familial Hypercholesterolemia

This condition results from a heritable lack of functional receptors for LDL. Since endocytosis of LDL cannot occur, intracellular
HMG CoA reductase becomes more active, while at the same time blood
levels of LDL reach very high levels (See Chapter 11). Diet has little effect on the homozygous condition; drugs can help somewhat.

VII. CHOLESTEROL IN PERSPECTIVE

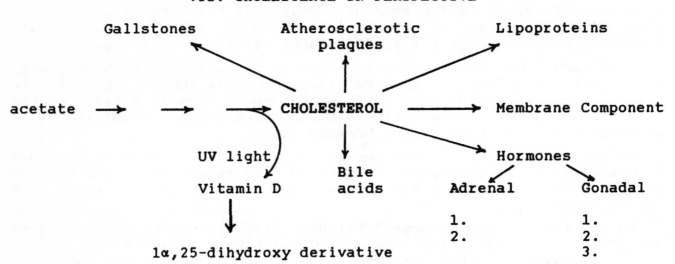

VIII. REVIEW QUESTIONS ON STEROIDS

DIRECTIONS: Each of the questions or incomplete statements below is followed by five suggested answers or completions. Select the one that is BEST in each case and fill in the corresponding space on the answer sheet.

1. In man,

A. all plasma cholesterol is carried as cholesterol esters.
B. cholesterol is made only by liver cells.
C. cholesterol in brain exchanges readily with cholesterol in plasma.
D. cholesterol is entirely metabolized to CO_2 and H_2O.
E. conversion of cholesterol to bile acids involves shortening the side chain.

2. Calcitriol is the most active form of

A. vitamin K
B. vitamin E
C. vitamin D
D. vitamin B_{12}
E. vitamin A

3. Estriol is excreted in the urine as the conjugate with

A. glucuronic acid
B. cysteine
C. glycine
D. glutamine
E. protein

4. Which of the following stimulates the synthesis of calcium binding protein in intestinal mucosa?

A. calcium
B. 1,25-dihydroxycholecalciferol
C. alpha-tocopherol
D. calcitonin
E. parathyroid hormone

5. In mammals, which of the following can NOT take place?

A. estrone $\longrightarrow$ estradiol
B. estrone $\longrightarrow$ dihydrotestosterone
C. 17-hydroxypregnenolone $\longrightarrow$ testosterone
D. progesterone $\longrightarrow$ estrogen
E. cholesterol $\longrightarrow$ estrogen

6. Serum cholesterol levels may be significantly changed by a diet with a low ratio of:

A. diglycerides to monoglycerides
B. saturated to polyunsaturated fatty acids
C. lactose to glucose
D. vitamin K to vitamin D
E. cephalin to lecithin

7. All of the following depend on micellar activity for their absorption EXCEPT

A. glycine
B. vitamin E
C. cholesterol
D. vitamin A
E. stearic acid

8. Some virilization may be seen in a patient with Cushing's syndrome, but not in a patient on high doses of prednisone because

A. prednisone has weak glucocorticoid activity and prolonged use does not cause adrenal atrophy
B. prednisone has weak mineralocorticoid activity and is a strong glucocorticoid
C. prednisone cannot produce a Cushingoid toxicity syndrome
D. virilization in the patient with Cushing's syndrome is caused by adrenal androgens
E. Cushing's syndrome is an autosomal dominant trait

DIRECTIONS: For each of the questions or incomplete statements below, ONE or MORE of the answers or completions is correct. On the answer sheet fill in space

A if only 1, 2, and 3 are correct
B if only 1 and 3 are correct
C if only 2 and 4 are correct
D if only 4 is correct
E if all are correct

FILL IN ONLY ONE SPACE ON YOUR ANSWER SHEET FOR EACH QUESTION

Directions Summarized				
(A) 1,2,3 only	(B) 1,3 only	(C) 2,4 only	(D) 4 only	(E) All are correct

9. Biosynthesis of cholesterol involves:

1. lanosterol
2. squalene
3. dimethylallyl pyrophosphate
4. succinyl CoA

10. In fasting there is lowered activity of liver HMG CoA reductase and squalene oxidocyclase, and therefore less synthesis of

1. HMG CoA
2. ubiquinone
3. acetoacetate
4. cholesterol

11. Cholesterol can be converted to

1. androgens
2. corticosteroids
3. bile acids
4. polyunsaturated fatty acids

12. In man, cholesterol

1. is converted to glycocholic acid by intestinal bacteria.
2. is a part of cell membranes
3. can be catabolized mainly to acetyl CoA.
4. is a precursor of cortisol.

13. Which of the following promotes mineralization of bone?

1. decreased levels of ascorbic acid
2. increased levels of plasma calcium
3. parathyroid hormone
4. 1,25-dihydroxycholecalciferol

14. Vitamin D

1. is not itself active in stimulating calcium transport by intestine and calcium mobilization from bone in vitro.
2. is produced by ultraviolet light acting on delta-5,7 sterols.
3. is converted to the 1,25-dihydroxy-derivative which stimulates the intestinal mucosa to transport calcium.
4. induces the kidney to excrete phosphate.

15. High doses of hydrocortisone over 2 months would probably cause

1. a rise in liver glycogen
2. larger skeletal muscles
3. atrophy of the adrenal cortex
4. a rise in blood ACTH

16. Agents that lower blood cholesterol include

1. cholestyramine
2. compactin
3. clofibrate
4. nicotinic acid

17. The mechanism of action of progesterone involves

1. interaction with a membrane-bound receptor
2. penetration of the hormone into the cell
3. activation of adenylate cyclase
4. synthesis of new protein

FILL IN ONLY ONE SPACE ON YOUR ANSWER SHEET FOR EACH QUESTION

Directions Summarized				
(A)	(B)	(C)	(D)	(E)
1,2,3	1,3	2,4	4	All are
only	only	only	only	correct

18. The biosynthesis of testosterone involves

1. isopentenyl pyrophosphate
2. pregnenolone
3. lanosterol
4. cortisol

19. A lack of receptors for LDL will probably induce

1. a low level of blood LDL
2. high activity of HMG CoA reductase
3. decreased plasma cholesterol
4. decreased endocytosis of LDL

20. The surge of gonadotropins around the time of ovulation

1. is caused by the effect of GnRH on the ovary
2. is caused by a positive feedback of estradiol secreted during the follicular phase
3. starts the development of follicles that release estradiol during the luteal phase
4. starts the development of the *corpus luteum* that releases progesterone during the luteal phase

IX. ANSWERS TO QUESTIONS ON STEROIDS

1. E
2. C
3. A
4. B
5. B
6. B
7. A
8. D
9. A
10. D

11. A
12. C
13. C
14. A
15. B
16. E
17. C
18. A
19. C
20. C

9. PURINES AND PYRIMIDINES

Leon Unger

Nucleic acids are comprised of nitrogenous bases (purines and pyrimidines), pentose sugars (ribose and deoxyribose) and phosphate groups. Specific sequences of purines and pyrimidines encode the genetic information of cells and organisms.

I. STRUCTURE AND NOMENCLATURE

A. Nitrogenous Bases

The purines, **adenine (A)** and **guanine (G)**, are present in both RNA and DNA. Catabolism of adenine and guanine produces the purines **inosine, hypoxanthine, xanthine** and **uric acid.**

The pyrimidines, **cytosine (C)** and **thymine (T)** are present in DNA. Cytosine and **uracil (U)** are contained in RNA (Figure 9.1).

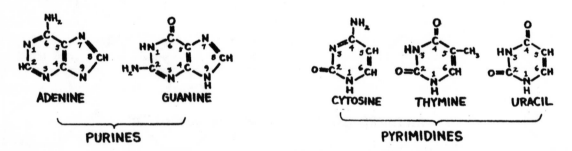

Figure 9.1. The Purine and Pyrimidine Bases.

B. Nucleosides

A **nucleoside** consists of a purine or pyrimidine base linked to a pentose sugar. In RNA, the pentose is **D-ribose**, therefore the nucleoside formed is termed a ribonucleoside. (example: adenosine). The pentose is **2-deoxy-D-ribose** in in DNA, and the nucleoside is called a deoxyribonucleoside. (example: deoxyadenosine). The atoms of the pentoses are designated by primed numbers to distinguish them from the numbers in the bases (Figure 9.2).

In a nucleoside, carbon 1' (C-1') of the pentose is linked by a ß-glycosidic bond to N-9 of the purine or N-1 of the pyrimidine.

N-9-purine   N-1-pyrimidine
| |
pentose-C-1'   pentose-C-1'

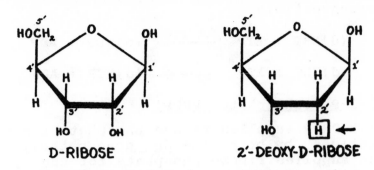

Figure 9.2. The Pentoses of Nucleic Acids

The major ribonucleosides for the purines are **adenosine** and **guano-sine** and for the pyrimidines, **cytidine** and **uridine**.

The major deoxyribonucleosides are **deoxyadenosine, deoxyguanosine, deoxycytidine** and **(deoxy)thymidine** (Figure 9.3).

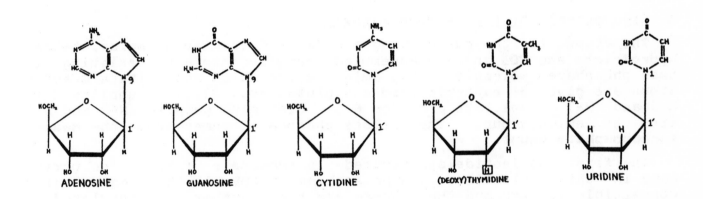

Figure 9.3. The Major Nucleosides

C. Nucleotides

A nucleotide is a nucleoside in which a phosphate group has been esterified to a hydroxyl group on the pentose. The most common position to be esterified is at C-5'. Other positions for esterification include the hydroxyls at C-3' (common) and C-2' (less common).

II. SYNTHESIS OF NUCLEOSIDE DIPHOSPHATES AND TRIPHOSPHATES

The ribo- or deoxyribo-nucleoside 5'-monophosphates [NMP,(d)NMP] may be further phosphorylated. Using ATP as a phosphate donor, <u>nucle-oside monophosphate kinase</u> phosphorylates (d)NMP to (d)NDP. <u>Nucleoside</u>

diphosphate kinase then converts (d)NDP to (d)NTP. These nucleotides are readily interconvertible.

$$(d)NMP + ATP \longrightarrow (d)NDP + ADP$$

$$(d)NDP + ATP \longrightarrow (d)NTP + ADP$$

5-Phosphoribosyl-1-Pyrophosphate (PRPP)

1. Functions: PRPP supplies ribose phosphate for the synthesis of purine ribonucleotides by both the de novo pathway and the salvage pathway. It also supplies ribose phosphate for the synthesis of pyrimidine ribonucleotides. PRPP is an intermediate in **histidine** and **tryptophan** biosynthesis.

2. Synthesis: PRPP is synthesized from ribose-5-phosphate and ATP. The enzyme catalyzing this reaction is ribose phosphate pyrophosphokinase (phosphoribosylpyrophosphate synthetase).

$$Ribose-5-P + ATP \longrightarrow PRPP + AMP$$

III. PURINE METABOLISM

A. Biosynthesis by the De Novo Pathway

The atoms for the purine ring are derived from **amino acids, tetrahydrofolate** and **CO_2** and are assembled stepwise onto a pre-existing ribose phosphate molecule. The amino acids which contribute C and N atoms are **glycine, aspartic acid** and **glutamine**. Glycine supplies C-4, C-5 and N-7. Aspartate provides N-1. Glutamine supplies N-3 and N-9 from the amido group. The ribose phosphate comes from the **hexose monophosphate shunt**.

The first nucleotide synthesized is **inosinate** (IMP). The purine base in this nucleotide is **hypoxanthine**. IMP is then sequentially convertible to AMP and GMP. These are then further phosphorylated to ATP and GTP.

The first step specific to the synthesis of IMP is the formation of **5-phosphoribosylamine** from PRPP and glutamine. In this reaction, mediated by amidophosphoribosyl transferase, the pyrophosphate group of PRPP is replaced by an amino group donated by glutamine. The irreversible formation of phosphoribosylamine commits PRPP to purine synthesis. The amidotransferase is inhibited by **azaserine**.

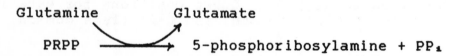

The next step involves the addition of the entire **glycine** molecule. The glycine atoms become C-4, C-5 and N-7 of the purine ring. This step requires the utilization of one ATP.

C-8 is provided by the formyl group of **N^5, N^{10}-methenyltetrahydrofolate**.

N-3 is added from the amide group of **glutamine**. This step utilizes an ATP.

The next step involves the cyclization of the molecule to form the five-membered **imidazole** ring.

C-6 is added by carboxylation with CO_2. **Biotin** is a cofactor in this process.

N-1 is provided from **aspartate**. Consumption of a molecule of ATP is required.

C-2 is donated by from N^{10}-formyltetrahydrofolic acid.

Then ring closure to form the first complete purine, IMP, or hypoxanthine ribose phosphate:

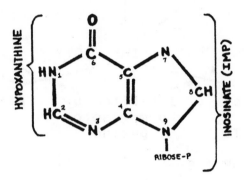

The origins of the atoms in the purine ring are summarized in Figure 9.4.

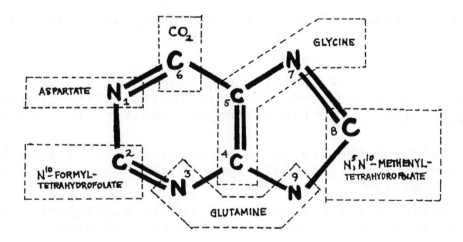

Figure 9.4. Origin of the Atoms in Purines

Finally, AMP and GMP are formed from IMP as shown in Figure 9.5.

B. Control of *De Novo* Purine Nucleotide Biosynthesis

There are three sites at which feedback inhibition regulates *de novo* purine nucleotide biosynthesis (Figure 9.5):

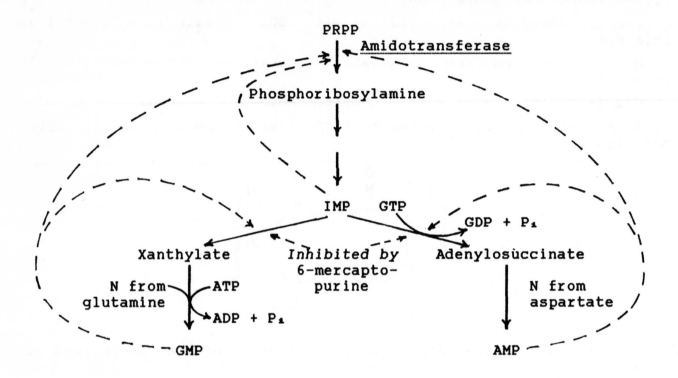

Figure 9.5. Synthesis of Adenylate and Guanylate, and Regulation by Nucleotides

1. Excess AMP and GMP inhibits the conversion of PRPP to phosphoribosylamine. These end-products synergistically inhibit <u>amidophosphoribosyl transferase</u>, the first enzyme specific for purine biosynthesis.

2. AMP inhibits the reaction converting IMP to **adenylosuccinate**, thereby preventing further AMP synthesis. This inhibition does not affect GMP formation.

3. GMP inhibits the conversion of IMP to **xanthylate** (XMP), preventing its own formation without affecting AMP synthesis.

In addition, there is a cross-regulation between the two branches of purine synthesis: ATP is required to convert xanthylate to GMP, and GTP is needed in the formation of adenylosuccinate from IMP.

C. Catabolism of Purine Nucleotides and Their Regeneration by the Salvage Pathway

During digestion, nucleic acids are cleaved to purine and pyrimidine nucleotides by the action of pancreatic ribonucleases and and deoxyribonucleases. These nucleotides are degraded to their corresponding nucleosides and then to the free bases. With respect to the purines, the bases may be 1) <u>salvaged</u> to regenerate the parent nucleotide or 2) further <u>catabolized</u> to urate (in man)(Figure 9.6).

The nucleotides AMP, IMP and GMP are hydrolytically cleaved by <u>purine 5'-nucleotidase</u> to the nucleosides adenosine, inosine and guanine, respectively. These nucleosides are catabolized to the free bases adenine, hypoxanthine and guanine by <u>purine nucleoside phosphorylase,</u> liberating ribose-1-phosphate. Instead of being converted to adenine, adenosine may be deaminated to inosine by <u>adenosine deaminase</u>.

One salvage enzyme, <u>adenine phosphoribosyl transferase (APRTase)</u>, catalyzes the transfer of ribosyl phosphate from PRPP to adenine to resynthesize AMP. A second salvage enzyme, hypoxanthine-guanine phosphoribosyl transferase (HGPRTase), mediates the transfer of ribose phosphate to hypoxanthine or guanine to regenerate IMP or GMP, respectively.

$$\text{Adenine + PRPP} \xrightarrow{(1)} \text{AMP + PP}_i$$

$$\text{Hypoxanthine + PRPP} \xrightarrow{(2)} \text{IMP + PP}_i$$

$$\text{Guanine + PRPP} \xrightarrow{(2)} \text{GMP + PP}_i$$

1 = adenine phosphoribosyl transferase (APRTase)
2 = hypoxanthine-guanine phosphoribosyl transferase (HGPRTase)

If hypoxanthine or guanine is not processed through the salvage pathway and recycled, it is oxidized to **xanthine** and then to **uric acid**; these two final steps are catalyzed by <u>xanthine oxidase</u>. **Uric acid** is the final principal end-product of purine catabolism in man and is excreted in the urine (Figure 9.6).

D. <u>Disorders of Purine Metabolism</u>

1. <u>Gout</u>: **Gout** is the result of the excessive production of uric acid and its subsequent deposition, as poorly soluble sodium urate crystals, in the joints and kidneys. The overproduction of urate results from the catabolism of abnormally high levels of purines synthesized *via* the *de novo* pathway. The high concentration of purines may be due to an increase in PRPP either because of 1) <u>decreased utilization</u> of PRPP by the salvage pathway caused by a partial deficiency of HGPRTase activity, or 2) <u>enhanced production</u> of PRPP because of overactivity of <u>phosphoribosyl pyrophosphate synthetase</u>. High levels of purines may also result from an increase in **phosphoribosylamine (PR-amine)** due to overactivity of <u>amidophosphoribosyl transferase</u> which is the rate-limiting enzyme in purine synthesis.

Gout is treated with **allopurinol**, an analog of hypoxanthine (Figure 9.6). Allopurinol inhibits xanthine oxidase and, therefore, prevents the conversion of hypoxanthine to xanthine and then xanthine to urate.

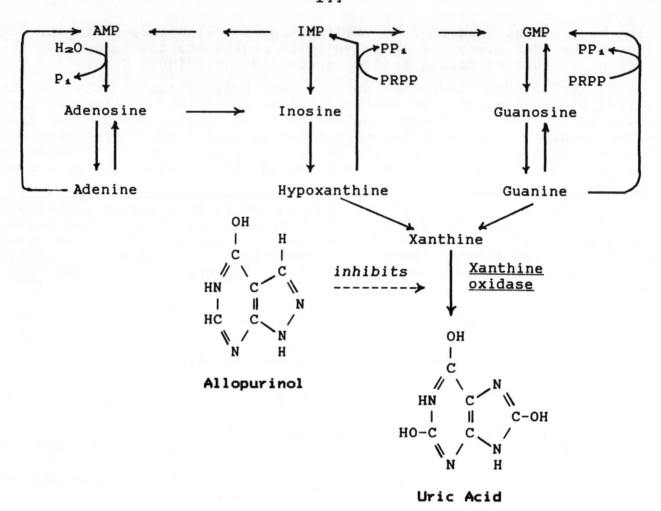

Figure 9.6. Catabolism and Salvage of Purine Nucleotides.

This results in a reduction of serum urate concentration and increases the concentrations of the more soluble hypoxanthine and xanthine.

2. <u>Lesch-Nyhan Syndrome</u>: The **Lesch-Nyhan Syndrome** results from a nearly complete absence of HGPRTase because of a genetic defect. Consequently, there is an absence or marked decrease of hypoxanthine and guanine salvage, an elevated PRPP concentration and an increased purine nucleotide synthesis by the <u>de novo</u> pathway. The ensuing catabolism of the over-produced purines leads to increased urate production. Victims of this disease show <u>gout</u>, <u>mental retardation</u>, <u>self-mutila-tion, hostility</u> and a <u>lack of muscular coordination</u>. Allopurinol is not effective in alleviating the symptoms of this disease. The syndrome is inherited as a <u>sex-linked recessive</u> trait in males.

3. <u>Immunodeficiency Diseases</u>

a. A deficiency of <u>adenosine deaminase</u> is associated with a se-vere combined deficiency of both T-cells and B-cells and deoxy-adenosinuria.

b. A <u>purine nucleoside phosphorylase</u> deficiency leads to a severe T-cell deficiency but normal B-cell function. Additional characteristics of this disorder include inosinuria, guanosinuria and <u>hypo-
uricemia</u>. Both of these disorders are inherited as <u>autosomal reces-
sive</u> traits.

IV. PYRIMIDINE METABOLISM

Biosynthesis

Whereas the growing purine ring is assembled on ribose-P, the pyrimidine ring is formed first and then joined to ribose-P to form a pyrimidine nucleotide. (The steps below are identified in Figure 9.7).

Step 1: The regulatory step in the synthesis of pyrimidines in
mammals is the formation of **carbamoyl phosphate** by cytoplasmic <u>carba-
moyl phosphate synthetase</u>. This enzyme is inhibited by UTP, a product
of the pathway.

Step 2: The committed step in the synthesis of the pyrimidine ring
in bacteria is the formation of **N-carbamoylaspartate** from carbamoyl
phosphate and aspartate. The carbamoyl phosphate atoms become C-2 and
N-3. Aspartate donates C-4, C-5, C-6 and N-1. The carbamoylation of
aspartate is catalyzed by <u>aspartate transcarbamoylase</u> (<u>ATC-ase</u>). In
bacteria, ATC-ase is the rate limiting enzyme; it is feedback-inhibi-
ted allosterically by CTP, the end-product of the pathway.

Steps 3 and 4: Ring closure and oxidation yields the pyrimidine,
orotic acid.

Step 5: Ribose-P, derived from PRPP, is attached to orotate to
yield the nucleotide, **orotidylate (orotidine-5-P)**.

Step 6: Orotidylate is decarboxylated to form **UMP**. All other pyrimidine nucleotides are synthesized from UMP.

Steps 7 and 8: UMP is phosphorylated sequentially to **UDP** and **UTP**.

Step 9. UTP is aminated by glutamine to form **CTP**.

V. DEOXYRIBONUCLEOTIDES ARE FORMED BY REDUCTION OF RIBONUCLEOSIDE DIPHOSPHATES.

Deoxyribonucleotides are formed from ribonucleotides. The ribose
of the ribonucleoside diphosphate is reduced at the 2' carbon atom to
form the 2'-deoxyribonucleoside diphosphate.

$$NDP \longrightarrow dNDP$$

For example, UDP $\longrightarrow$ dUDP

DNA also differs from RNA in that it contains **thymine** instead of
uracil. Thymine is uracil methylated at the C-5 position. **Deoxythymidylate (dTMP)** is produced by the methylation of deoxyuridylate
(dUMP) by tetrahydrofolate (Figure 9.7).

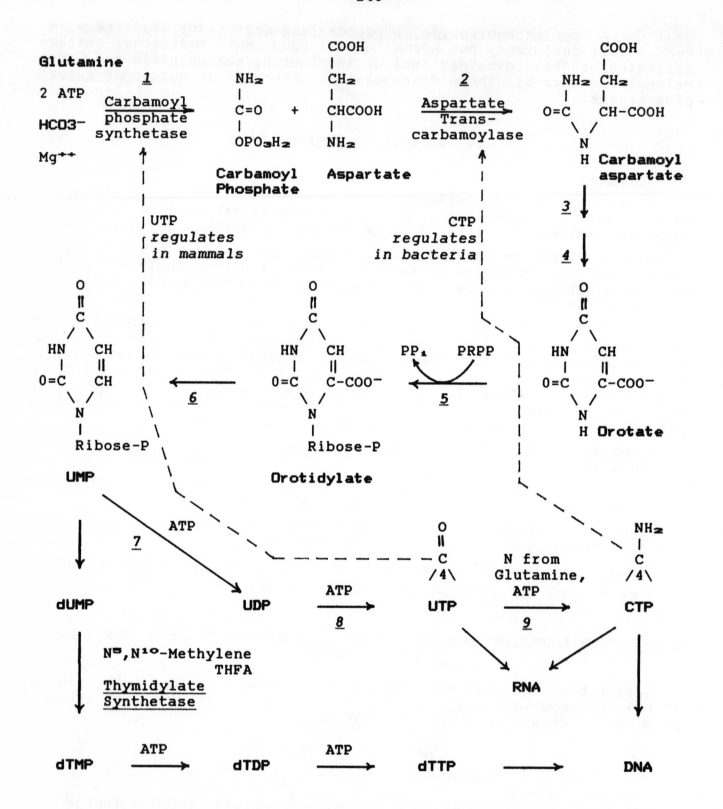

Figure 9.7. Biosynthesis of Pyrimidine Nucleotides, and Sites of Feedback Control

VI. SOME ANTICANCER DRUGS ACT BY BLOCKING DEOXYTHYMIDYLATE SYNTHESIS

Several drugs are used to prevent the conversion of dUMP to dTMP which is required for DNA synthesis. **5-Fluorouracil (5-FU)** irreversibly inhibits <u>thymidylate synthetase</u> (Figure 9.8) and is particularly useful in treating solid tumors. The folic acid antagonists, **aminopterin** and **amethopterin (methotrexate)**, inhibit <u>dihydrofolate reductase</u>, preventing the regeneration of tetrahydrofolate and inhibiting dTMP synthesis. Methotrexate is used to treat leukemia and choriocarcinoma.

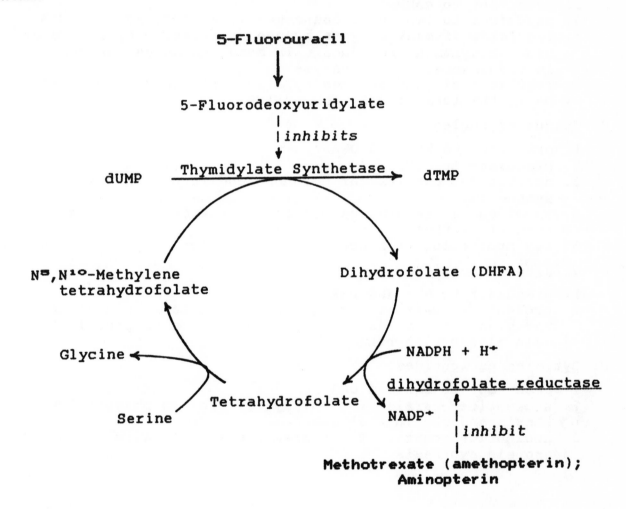

Figure 9.8. Anticancer Drugs which Block the Synthesis of Deoxythymidylate.

VII. FUNCTIONS OF NUCLEOTIDES

The nucleoside diphosphates and triphosphates are the "active" forms of the nucleotides. Some functions of the nucleotides are:

A. Adenine nucleotides

1. immediate energy source for most enzymatic reactions requiring energy expenditure. The form in which most biological energy is stored.
2. precursor to RNA and DNA.
3. precursor to cAMP
4. precursor to important coenzymes like NAD^+, FAD and CoA.
5. regulator of many enzymes by phosphorylation of serine or threonine or tyrosine residues. In some instances, an adenosyl group is transferred to the enzyme.
6. regulator of many enzymes by allosteric mechanisms; effector may be in the form of ATP, ADP or AMP.

B. Guanosine nucleotides

1. precursor to RNA and DNA.
2. precursor to cGMP
3. nucleotide carrier for mannose and fucose in glycoprotein biosynthesis.
4. involved in several key steps in peptide bond formation in protein biosynthesis.
5. key nucleotide in biochemistry of vision.

C. Uridine nucleotides

1. precursor to RNA and DNA
2. nucleotide carrier for glucose, galactose, N-acetylglucosamine and glucuronic acids in polysaccharide, glycoprotein and glycosaminoglycan biosyntheses.

D. Cytidine nucleotides

1. precursor to RNA and DNA
2. nucleotide carrier for diacylglycerol in phosphoglyceride synthesis and choline salvage.
3. nucleotide carrier for neuraminic acid (sialic acids) in glycoprotein synthesis.

10. NUCLEIC ACIDS AND PROTEIN SYNTHESIS

Jay S. Hanas

I. NUCLEIC ACIDS: STRUCTURE, SYNTHESIS, FUNCTION

A. DNA Structure

Deoxyribonucleic acid (DNA) and ribonucleic acid (RNA) are responsible for the transmission of genetic information and are intimately involved in cellular metabolism, growth, and differentiation. The central dogma of Molecular Biology states that genetic information is transferred from DNA to RNA to protein.

$$DNA \longrightarrow RNA \longrightarrow Protein$$

DNA is a polymer of deoxynucleoside monophosphates (deoxynucleotides). The 5' phosphate group of one deoxynucleotide is joined to the 3'OH group of another, forming a phosphodiester linkage. A polymer of nucleotides (polynucleotide) results when many nucleotides are joined in linear fashion. In deoxyribonucleic acid, the nucleotide bases (**adenine, guanine, cytosine,** and **thymine**) of one strand form hydrogen bonds with the nucleotide bases of the other strand. (See Chapter 9 for structures of these compounds.) Figure 10.1 shows two DNA strands linked together in a double helix by hydrogen bonding between the bases. Those bases linked by three hydrogen bonds are GC base pairs and those by two bonds are AT base pairs.

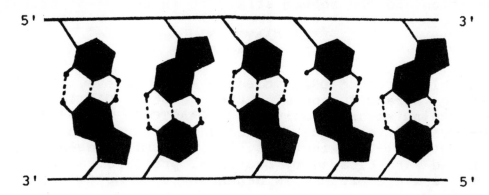

<u>Figure 10.1</u>. Pairing of Bases in Double-stranded DNA

In the double stranded structure, the DNA strands are **anti-parallel,** i.e. one strand is in the 5'-3' orientation; the other strand is in the opposite 3'-5'orientation. DNA can also exist in single-stranded forms. DNA molecules can contain millions of nucleotides. Genetic information is stored as unique sequences of the four nucleotide bases. The **genome** of an organism consists of the total DNA content in base pairs in a cell nucleus. The human genome is considerably smaller than that of some plants and amphibians. The significance of differing genome sizes is not understood.

Double-stranded DNA forms a double helix structure in which the two strands are intertwined as two right-handed helices (the helices in DNA turn clockwise as one looks up the center of the molecule). One turn of the double helix represents 10.4 base pairs in the "B" form of DNA. The two strands of a duplex DNA molecule come apart (breaking of hydrogen bonds) and unwind in a process called **denaturation**. This is accomplished by high temperature, alkali, or specific enzymes. The T_m of DNA is the temperature at which 50% of the bases become unpaired and is proportional to the GC content. The individual strands can re-anneal in a process called **renaturation** if the DNA is slowly cooled, or the pH brought to neutrality.

The ability of complementary DNA sequences to reanneal or **hybridize** is the basis for identifying specific DNA sequences of clinical inter-est. For example, the presence of a particular virus in blood can be detected with a radiolabeled, single-stranded DNA probe which hybridi-zes only to the viral genome and not to lymphocyte DNA. The assay can be made very specific and very sensitive.

The individual nucleotide bases in double-stranded nucleic acids absorb less UV light than when in single-stranded molecules or the free base, a phenomenon called **hypochromicity**. When DNA is denatured, UV absorbance of the sample increases (the **hyperchromic shift**). Sin-gle-stranded nucleic acids can form intrastrand base-pairs exhibiting partial double stranded character.

By convention, the top strand of any printed DNA sequence is in the 5'-3' orientation and the bottom strand is in the 3'-5' orientation.

$$5'-GAATTC-3'$$
$$3'-CTTAAG-5'$$

In the double-stranded DNA molecule shown above, the top strand (the non-coding strand) contains the sequence GAATTC. The bottom strand (the coding strand) also reads GAATTC in its 5'-3' direction. This double-stranded sequence is called a **palindrome** because it reads the same in both 5'-3' directions. Palindromic sequences in DNA are often recognized and cleaved by bacterial proteins called restriction enzymes. The cleavage process is dependent upon Mg^{++} ions.

The sequence above can be cut by the restriction enzyme **EcoRI** (de-rived from the intestinal bacterium Escherichia coli) at the following bases:

```
GAATTC    EcoRI     GAATTC                G       +   AATTC
CTTAAG    Mg++      CTTAAG    ------>     CTTAA        G
```

The enzyme DNA ligase can then covalently rejoin the cleaved pieces of DNA in a process dependent upon Mg and ATP as follows:

```
G       +   AATTC    DNA ligase     GAATTC
CTTAA        G       ATP, Mg++      CTTAAG
```

The use of restriction enzymes to specifically cut double-stranded DNA and of ligases to rejoin DNA molecules at specific sites forms the basis of recombinant DNA technology.

B. <u>DNA Replication</u>

DNA replication results in the exact duplication of the double-stranded DNA molecule. DNA replication takes place by a **semiconservative** pathway in which the two strands of duplex DNA are separated and each single strand serves as a template for the synthesis of the second, complementary strand (Figure 10.2).

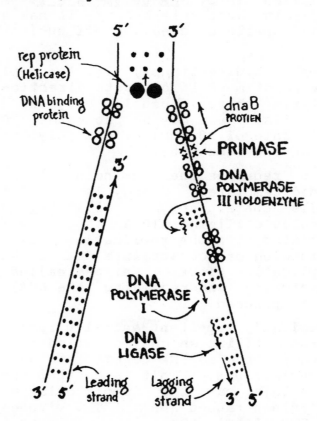

<u>Figure 10.2</u>. Scheme for DNA Replication in Procaryotes

The process of DNA synthesis is better understood in bacteria than in mammalian cells. The bacterium *E. coli* contains three distinct <u>DNA polymerases</u> (the enzyme which catalyzes the polymerization of deoxynucleotides on the template strand), called **DNA pol I, II,** and **III**. Mammalian cells contain at least two DNA polymerases, a nuclear enzyme and a distinct mitochondrial enzyme. DNA polymerases require a free 3'OH group for reaction with an incoming deoxynucleotide 5' phosphate. Thus, the template strand needs a **hybridized primer** to <u>initiate</u> DNA synthesis. The primer is usually a small RNA strand (RNA polymerase does not need a primer to begin synthesis). DNA pol I in bacteria is the most abundant DNA polymerase and is used for repair synthesis of DNA.

DNA synthesis is continuous on one strand (**leading strand**) and discontinuous on the other (**lagging strand**). This means one strand of the parental DNA duplex is copied at the replication fork by the DNA polymerase in a continuous fashion. Since new nucleotides can be attached only to the 3'OH of the growing chain, the other strand needs a "back-stitching" process to form the new strand. Primers are of continual necessity for this back-stitching process (Figure 10.2). The discontinuous nature of DNA synthesis results in the production of newly synthesized DNA fragments called **Okazaki fragments**. DNA ligase joins the Okazaki pieces once the RNA primer has been degraded and the gaps have been filled in by DNA polymerase I. DNA polymerase I also can correct errors in synthesis because it has a 3'to 5' exonuclease activity which is capable of cleaving off nucleotides which were misincorporated.

DNA replication is **bidirectional**, i.e. two replication forks initiated at the same point move in opposite directions on the same duplex strand. In bacteria there is usually only one origin of replication but in mammalian cells, there may be thousands. DNA replication takes place at about the rate of 20 nucleotides polymerized per second. RNA synthesis occurs at the same rate.

DNA replication requires the unwinding of the duplex DNA to yield the single-stranded templates. Energy is needed to unwind or denature duplex DNA. Cells contain enzymes called helicases which unwind the duplex DNA molecules utilizing ATP as an energy source. This unwinding process associated with DNA replication would eventually come to a halt because of topological constraints due to the build up of "knots" ahead of the replication fork. Enzymes called topoisomerases relax these knots (by nicking and rejoining the DNA) and thus permit the replication fork to proceed.

A number of medically important animal viruses, e.g. **HTLV (human T-lymphotrophic virus)** utilize an enzyme called reverse transcriptase to convert their single-stranded RNA genome into duplex DNA genomes which integrate into nuclear DNA. This is another example of DNA replication, although in this case the template is a single-stranded RNA rather than a single-stranded DNA. HTLV viruses cause some forms of human leukemia. HTLV-III is the causative agent in **AIDS (acquired immune deficiency syndrome)**.

Deoxyribonucleoproteins: In most cases, nucleic acids in animal cells do not exist as free entities but are found complexed with specific proteins. The protein components protect the nucleic acids from non-specific degradation by nucleases. Key enzymes and other protein factors also bind to specific base sequences in the nucleic acids.

The DNA in the nucleus of eucaryotic cells is complexed tightly with basic proteins called **histones** which contain large amounts of **lysine** and **arginine**. There are five histones designated **H1, H2A, H2B, H3,** and **H4**. Histones interact with nuclear DNA in a very ordered way in the form of histone octomers composed of dimers of H2A, H2B, H3, H4. These structures are termed **nucleosomes**. About 145 base pairs of duplex DNA are wrapped around each nucleosome (slightly less than two turns) (Figure 10.3).

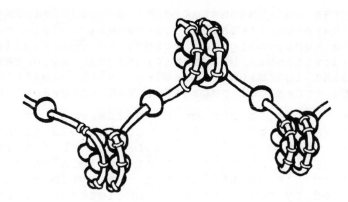

Figure 10.3. DNA and Histone Interactions to Form Nucleosomes

H1 is located where DNA enters and leaves the nucleosome. There exist 25-100 bp of **spacer** or **linker** DNA in the region between adjacent nucleosomes.

C. Chromatin and Chromosomes

In eucaryotic cells, nuclear DNA is packaged into nucleosomes along with many other non-histone proteins to form a nucleoprotein complex called **chromatin**. If chromatin is washed extensively with buffers containing high salt concentrations, all proteins are extracted except the core histones, H2A, H2B, H3, H4. The structures remaining resemble "beads" of nucleosomes on a "string" of DNA in the electron microscope. These "beads on a string" have diameters of about 10 nm. When chromatin is washed less extensively with salt, the structure observed in the electron microscope has a diameter of about 30 nm and appears to be composed of a helical array of nucleosomes.

DNA exists in the nuclei of eucaryotic cells as components of complex chromatin structures called **chromosomes**. These are highly condensed structures of the 30 nm chromatin fibers. The chromatin in chromosomes exists as **euchromatin** which is not so highly condensed and **heterochromatin** which is highly condensed. Human somatic cells (non-germ line) contain 22 pairs of non-identical chromosomes and 2 sex chromosomes (XX, female and XY male).

When chromosomes in the nucleus of eucaryotic cells are replicated, the two DNA strands are separated and new complementary strands are synthesized on each separated strand. The nucleosome structure must also be replicated as well. All the "old" nucleosomes remain associated with one of the separated strands and new nucleosomes (composed of newly synthesized histones from the cytoplasm) form on the other strand.

D. DNA Mutations

A **gene mutation** occurs when the unique sequence of deoxyribonucleotides in DNA is altered in any way. A mutation in the DNA sequence that codes for a protein can alter its amino acid sequence.

The consequences of DNA mutations on cellular metabolism can range from extremely harmful (lethal) to beneficial. Mutations can occur spontaneously due enzymological errors in DNA replication, recombination, and cell division. DNA mutations can also result from environmental factors like ionizing radiation, ultraviolet light from the sun and other sources, chemical mutagens, and viruses and viral genes.

Mutations can be of a number of different types. Point mutations involve the alteration of a single nucleotide or nucleotide pair. A **transition** occurs when a purine is substituted for another purine or a pyrimidine is substituted for another pyrimidine. A **transversion** occurs when a purine is substituted for a pyrimidine or *vice versa*. A **deletion** is caused by the loss of a nucleotide or nucleotide pair and an **insertion** is the gain of a nucleotide or nucleotide pair. **Gross mutations** involve more than one nucleotide pair and may include very large segments of the genome. A **translocation** results from the movement of a chromosomal segment to a non-homologous chromosome. **Inversions** are inverted pieces of DNA within one chromosome. Large deletions and insertions of DNA are also possible. Mutations which lead to defective proteins and metabolic alterations are usually deleterious to the cell. Some mutations can be harmless, leading to an amino acid change in a protein which has no effect on structure and function. Mutations can be beneficial if the mutated gene product confers a selective advantage to the organism. Table 10.1 is a partial list of mutagens and the types of mutations they cause.

The mutation rate in organisms due to ultraviolet and ionizing radiation should be be much higher than that actually observed. The low mutation rate is due in large part to the ability of organisms to repair DNA damage. **Thymine dimers**, induced by UV irradiation, can be repaired by photoreactivation, and involves the light-activation of specific enzymes (occurs in bacteria). T-T dimers are also removed by excision repair, in which an <u>endonuclease</u> (a nuclease which cleaves the phosphodiester bond between adjacent nucleotides) clips the phosphodiester backbone near the dimer, an <u>exonuclease</u> (a nuclease which cleaves phosphodiester bonds successively from an end) removes the dimer and surrounding nucleotides, and the gap is filled in by the action of <u>DNA polymerase I</u> and <u>DNA ligase</u> (Figure 10.4).

T-T dimers can also be repaired by recombination events. In persons with **Xeroderma pigmentosum**, defective excision-repair enzymes make these individuals prone to skin cancers.

E. <u>RNA Structure</u>

RNA is usually single-stranded and contains rarely more than a few thousand nucleotides joined in phosphodiester linkages. Compared to DNA, cellular RNA is much more varied in size, structure, and function. Animal cells contain **heterogeneous nuclear RNA (hnRNA), messenger RNA (mRNA), small nuclear RNA (snRNA), ribosomal RNA (rRNA), transfer RNA (tRNA)** and other less well understood species. The majority of the RNA in cells is rRNA (80%), followed by tRNA (5%), mRNA (2%), and other (3%). RNA is hydrolyzed by alkali because of the presence of adjacent 2',3'-OH groups on the ribose whereas DNA (contain-

Mutagen	Mutation Type
5-Bromouracil	transitions, replaces thymine in DNA and, after isomerization, pairs with G instead of A
aminopurine	transitions
hydroxylamine	transitions, alters C so it pairs with A
nitrous acid	transitions, deaminates cytosine to uracil which pairs with A rather than G
sulfur mustards nitrogen mustards ethylethane sulfonate methylmethane sulfonate	transitions and transversions by alkylating N7 of G, removes purines (apurination)
low pH	transitions and transversions, causes apurination
intercalating dyes (proflavine, 5-amino acridine, acridine orange, ethidium bromide)	insertions and deletions, causing frameshift mutations
ultraviolet (UV) light	dimerizes adjacent pyrimidines (thymine dimers) leading to deletions
ionizing radiation (X-rays, radioactivity, cosmic rays)	severe damage to DNA by breaking covalent bonds

Table 10.1. Mutagens and their Effects.

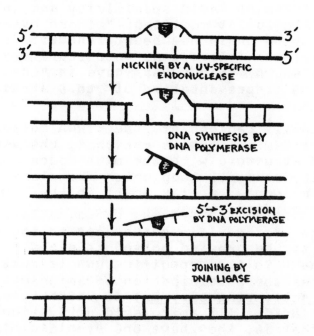

Figure 10.4. Mechanism of DNA Repair: Excision of Thymine Dimers.

ing only the 3'OH group) is denatured but not hydrolyzed by alkali. RNA can contain unusual (minor) nucleotide bases like **pseudouracil, ribothymidine, methyladenine,** and **dihydrouracil.**

1. <u>Messenger RNA</u>: Messenger RNA is synthesized in the nucleus of eucaryotic cells. The mixture of primary transcripts which act as precursors to mRNA is referred to as heterogeneous nuclear RNA. **Heterogeneous nuclear RNA (HnRNA)** is very unstable and can be the largest of RNA species (10-20,000 bases). It is processed rapidly into mRNA in the nucleus.

Messenger RNA (mRNA) which carries the genetic information from DNA to the ribosomes is found primarily in the cytoplasm. Eucaryotic mRNA has an unusual structure at its 5' end called a **"cap"** and has the following structure.

```
              5'  5'
     7-methyl GpppG————————mRNA————————AAAAAAAAAAAA
        "cap"                           "poly(A) tail"
```

It is linked 5'OH to 5'OH at the 5' end of the mRNA. Up to 200 A's may be attached to the 3' end of mRNA. This structure is called the **poly(A) "tail"** and has been useful in the isolation of animal mRNA by hybridization on columns containing long polymers of T's. Messenger RNAs are usually no more than several thousand bases in length.

Messenger RNAs in bacteria can be **polycistronic** (coding for more than one protein). The mRNAs in eucaryotic organisms tend to be **monocistronic** (coding for only one protein).

2. <u>Transfer RNA</u>: **Transfer RNAs (tRNA)** are one of the smallest RNA species found in the cell with lengths varying between 70-90 nucleotides. They are similar in size and structure, but have differences in base composition, in amino acid specificity and in codon recognition ability. They contain 10-20% modified, "minor" bases which are formed <u>after</u> the polynucleotide chain has been transcribed. All tRNAs can be folded into the familiar "cloverleaf" structure. X-ray crystallographic analyses have shown that tRNAs have an "L"-shaped three dimensional structure. A representation of this structure is shown in Figure 10.5.

Important structural regions of the tTRNA molecule are the CCA 3' terminus to which an amino acid is attached, the anticodon loop which forms a base-paired structure with the mRNA codon on the ribosome, the TUC loop and the dihydrouracil loop which are close together in the 3-D structure of the molecule but far apart in the 2-D structure. These loops contain many modified bases.

Each tRNA carries its specific amino acid to the surface of the ribosome to be added to the growing polypeptide chain. Before the amino acid can be attached to its specific tRNA it must first be "activated". This requires the participation of <u>aminoacyl synthetases</u> (also called <u>amino acid activating enzymes</u>) of which there are at least 20, one for each amino acid. All amino acid activating enzymes have **double specificity,** that is, they have one specific binding site for the amino acid and one for one or more specific tRNAs.

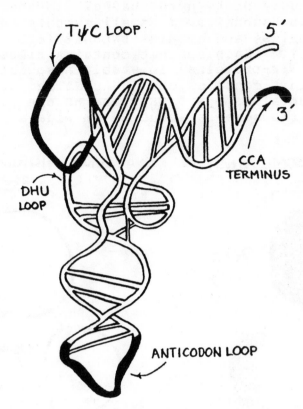

Figure 10.5. Three-dimensional structure of a representative tRNA molecule.

The amino acid is attached to the tRNA by an ester linkage between the α-carboxyl group of the amino acid and the 3'OH group of the terminal adenosine of the tRNA. All functional tRNAs end in -**CCA** at the 3' end.

The reactions catalyzed by these activating enzymes are reversible *in vitro* and can be depicted as follows.

$$(a) \qquad Enz_1 + ATP + AA_1 \rightleftharpoons Enz_1\text{-}AA_1\text{\textasciitilde}AMP + PP_1$$

$$(b) \quad Enz_1\text{-}AA_1\text{\textasciitilde}AMP + tRNA_1 \rightleftharpoons AA_1\text{-}tRNA_1 + Enz_1 + AMP$$

3. Ribosomal RNA and Ribosomes: The ribosomal RNAs are the major nucleic acid components of ribosomes. There are three major species of ribosomal RNA in **procaryotic** cells, a **5S, 16S** and **23S**.

The ribosomal RNAs of **eucaryotic** cells are composed of four species, the **5S, 5.8S, 18S** and **28S** RNAs. The 5.8S, 18S and 28S rRNAs are synthesized in the **nucleoli**. The 5S RNA is synthesized in the nucleus but not in the nucleolus. Ribosomal rRNAs contain some methylated adenine bases. Ribosomal RNAs, although single-stranded, exhibit a great deal of intrastrand base pairing resulting in highly ordered structures.

The major ribonucleoprotein in the cell is the ribosome. Ribosomes are composed of two subunits, a small subunit (40S in animal cells, 30S in bacterial cells) and a large subunit (60S in animal cells, 50S in bacterial cells). Each subunit contains ribosomal RNA (rRNA) and ribosomal proteins (r-proteins) in stable association. In eucaryotic cells, the 40S subunit contains one 18S rRNA and about 30 proteins and the 60S subunit contains one 5.8S, one 5S one 28S rRNA and about 40 proteins. The 80S ribosome is about 20 nm wide and 30 nm long (Figure 10.6).

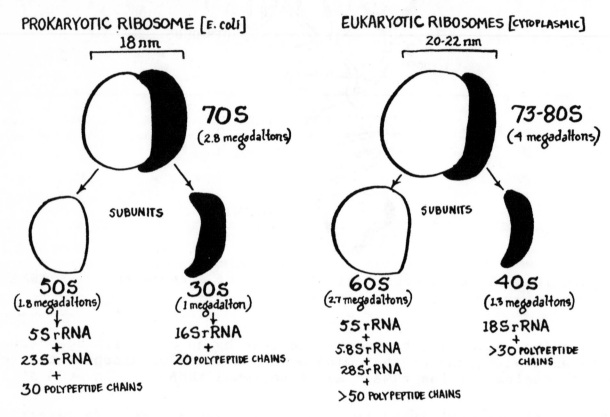

Figure 10.6. Composition of Procaryotic and Eucaryotic Ribosomes.

The free 80S ribosome is a monosome. **A polysome** is a mRNA strand complexed with up to 20 monosomes in the cytoplasm of animal cells. *All protein synthesis takes place on polysomes.* Free polysomes synthesis proteins destined to be used intracellularly. Membrane-bound polysomes synthesize proteins destined for secretion from the cell or destined to become membrane components. Mitochondria contain their own genetic system which includes a circular chromosome and mitochondrial ribosomes. Mitochondrial ribosomes are smaller than cytoplasmic ribosomes, having a sedimentation constant of 55S. Biologically the protein synthesizing machinery of mitochondria behaves as if it were procaryotic in nature.

4. Small Nuclear RNA and Small Cytoplasmic RNA: In procaryotic cells the coding regions of the DNA usually form one long continuous

sequence.   In eucaryotic cells, however, the coding regions are usually "split", that is, coding sequences (**exons**) are separated from one
another by non-coding sequences (**introns**) (Figure 10.7).   The functions of these introns are unknown. When the gene is transcribed, the
primary RNA transcript contains the sequences from both the exons and
the introns.   The introns must be removed and the exons "spliced" together to form the functional mRNA.   This splicing process is carried
out by ribonucleoprotein particles called **small nuclear ribonucleoprotein particles (snRNPs)** or **spliceosomes**.   The RNA components of these
particles are small, stable RNAs designated designated U1, U2,....,
ranging in size from 150-300 bases.   Several of these RNAs appear to
be involved in the processes converting hnRNA to mRNA.

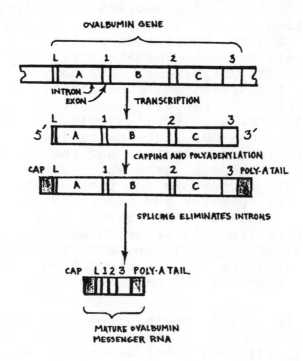

Figure 10.7. Structure of the Ovalbumin Gene

Patients with the autoimmune disease **systemic lupus erythematosus
(SLE)** and other autoimmune diseases produce autoantibodies against
these snRNPs as well as against other cellular nucleoproteins. Certain mutations in human globin genes will not allow the proper splicing of primary transcripts and are the cause of **thalassemias**.

An important **small cytoplasmic RNA (scRNA)** is the 7S RNA forming
part of the structure of the **signal recognition particle (SRP)**. The
SRP contains a 7S RNA (about 300 nucleotides in length) and 6-8 proteins.   This complex binds to the signal sequence at the amino terminal of secretory proteins and is involved in the secretion of nascent
proteins through the membranes of the RER.   See Section III A and
Figure 10.10.

F. RNA Synthesis (Transcription)

RNA synthesis is the process by which single-stranded RNA is synthesized from ribonucleotide triphosphate precursors by RNA polymerases utilizing single-stranded DNA as a template. The coding strand of the duplex DNA molecule is copied to form a 5'-3' RNA **transcript**. (Transcription as well as DNA polymerization is always in the 5'-3' direction.)

In bacteria, transcription factors include **sigma factor, catabolite activator protein (CAP)**, and **lactose operon repressor**. Although the E.coli genome contains only one site of initiation for DNA synthesis, thousands of RNA synthesis initiation sites exist, each containing specific DNA sequences called **promoters** which direct RNA polymerase to the proper initiation sites. In bacteria, this promoter is characterized by a "**TATA**" box at nucleotide position **-10** from the start of transcription (at +1) and also important sequences at nucleotide position **-35**. Unlike DNA polymerase, RNA polymerase does not require a primer to initiate transcription. The first nucleotide transcribed is a purine. Transcription factors regulate the initiation of RNA synthesis in bacterial cells. For example, the lactose **repressor** is a protein which binds to a DNA sequence called an "**operator**" which lies between the transcription promoter and the transcription initiation site for the ß-galactosidase gene which produces an enzyme which degrades lactose. If lactose is not present in the cell, then the repressor binds to the operator and blocks RNA polymerase from initiating transcription. If lactose is present in the cell, the sugar binds to repressor, alters its conformation so that it no longer binds to the operator. RNA polymerase is now able to transcribe the galactosidase gene. This results in the production of ß-galactosidase that degrades lactose, allowing the cell to use the breakdown products for metabolism.

RNA synthesis in animal cells appears to be more complicated than in bacteria, possibly because there are three distinct RNA polymerases, **RNA pol I, II, III**, and other associated RNA processing events.

RNA pol I transcribes the 45S precursor rRNA in the nucleolus. **RNA pol II** transcribes mRNA-coding genes. RNA pol II also transcribes the majority of the snRNAs. **RNA pol III** transcribes the 5S rRNA genes and tRNA genes and also the 7S RNA found in the signal recognition particle. The various eucaryotic RNA polymerases exhibit different sensitivities to the mushroom-derived toxin, **α amanitin**. RNA pol I is resistant to this toxin; RNA pol II is very sensitive and RNA pol III is moderately sensitive.

Ribosomal proteins synthesized in the cytoplasm enter the nucleus and migrate to the nucleoli where they assemble on the rRNA. 5S rRNA synthesized outside of the nucleolus also migrates to the nucleolus and becomes a component of the large (60S) subunit. The ribosomal subunits then migrate to the cytoplasm to form polysomes and take part in protein synthesis.

II. PROTEIN BIOSYNTHESIS

A. The Genetic Code

There are 20 amino acids commonly found in proteins which are linear arrays of amino acids joined by reacting the α-amino group of one
amino acid with the α-carboxyl group of the next amino acid. Given
four unique bases (A, U, G. or C), a singlet or doublet genetic code
would only give 4 or 8 unique combinations (4^1 or 4^2), clearly not
enough to code for 20 unique amino acids. A triplet code (4^3) would
code for 64 unique possibilities, more than enough for the 20 unique
amino acids. By using synthetic ribonucleotide polymers in crude *in
vitro* protein synthesizing systems the genetic code was indeed found
to be triplet in nature and codons (three adjacent ribonucleotides)
were assigned to all 20 amino acids (Table 10.2.

1st pos. 5'end	2nd position				3rd pos. 3'end
	U	C	A	G	
U	Phe	Ser	Tyr	Cys	U
	Phe	Ser	Tyr	Cys	C
	Leu	Ser	**Stop**	**Stop**	A
	Leu	Ser	**Stop**	Trp	G
C	Leu	Pro	His	Arg	U
	Leu	Pro	His	Arg	C
	Leu	Pro	Gln	Arg	A
	Leu	Pro	Gln	Arg	G
A	Ile	Thr	Asn	Ser	U
	Ile	Thr	Asn	Ser	C
	Ile	Thr	Lys	Arg	A
	Met	Thr	Lys	Arg	G
G	Val	Ala	Asp	Gly	U
	Val	Ala	Asp	Gly	C
	Val	Ala	Glu	Gly	A
	Val	Ala	Glu	Gly	G

Table 10.2. The Genetic Code

Note that the codons are presented as they would be found in mRNA.
Bolded codons indicate the initiation and chain terminations codons.

The code is **degenerate**; in many cases, more than one triplet codes
for the same amino acid. The code is **non-overlapping**; there are no
punctuation signals separating one codon from the next. The mRNA is
translated by the ribosomes, three nucleotides at a time, from a fixed

starting point in a single direction, 5' to 3'. This initiation point designates the reading "phase" or "frame". If a mutation occurs in a gene resulting in a mRNA with a nucleotide deleted or inserted, the reading frame will be altered (a **frame-shift mutation**). Once changed, the reading frame will not change unless a second, corrective mutation occurs "downstream" from the first. With one exception in mitochondria, the code is **universal**, that is, the same codon assignments designate a particular amino acid in all living things. Between species, however, each codon may be used with different frequencies.

Three codons do not code for amino acids. **These are UAA, UAG**, and **UGA** and are called the **termination** codons. They are also referred to as **stop** codons or **nonsense** codons. One codon, **UAG**, is the **initiation** codon. It codes for methionine and can also serve as an elongation codon. Its location within the mRNA determines its function.

B. Translation

The genetic code allows the translation of genetic information in mRNA (transcribed form of unique sequences of DNA) into amino acid sequences. Ribosomes, composed of ribosomal proteins and rRNAs, are the sites of cellular protein synthesis. The biosynthesis of proteins occurs by similar mechanisms in bacteria and mammalian cells. Protein synthesis is composed of three steps, **initiation, elongation**, and **termination**. The initiation step of protein synthesis positions the initiator codon of the mRNA, **AUG**, (sometimes GUG in procaryotes) and the initiator met-tRNA at the proper site on the ribosome for the start of translation. The 5'end of the mRNA in bacteria has a sequence which is complementary to the 3'end of the 16S RNA in the 30S subunit. Base pairing occurs between these complementary regions and assists in the proper positioning of the mRNA on the 30S subunit. Mammalian ribosomes do not contain this sequence homology. Mammalian mRNA has a 5'cap structure which may play a role in initiation of translation. Usually the first AUG from the 5'end of the mRNA serves as the initiating codon. In bacteria, the mRNAs are largely polycistronic so the ribosome needs to initiate at internal AUGs.

Bacteria and higher organisms initiate protein synthesis with the amino acid **methionine** (AUG codon). There are two met-tRNA species, one used only for initiation of protein synthesis and the other for elongation. Methionine, when aminoacylated to the initiator met-tRNA, is <u>formylated in bacteria but not in higher organisms</u>. Aminoacylation of tRNAs is catalyzed by enzymes called <u>tRNA synthetases</u>. At least one synthetase exists for each amino acid. The first step in tRNA aminoacylation is the formation of an aminoacyl-adenylate from the reaction of the amino acid with ATP. The aminoacyl moiety is then transferred to the 3'OH of the tRNA A which is always adjacent to two C's giving the CCA 3' terminus found in all tRNAs. This CCA is added posttranscriptionally as it is not coded in the tRNA gene.

Protein factors catalyze the various steps of protein synthesis. In bacteria, the initiation factors are designated **IF1, IF2**, and **IF3**. There are similar factors in eucaryotic cells. IF3 dissociates the 70S ribosome into the 30S and 50S subunits. The three initiation factors, fmet-tRNA, GTP, and mRNA then associate with the 30S subunit (Figure 10.8).

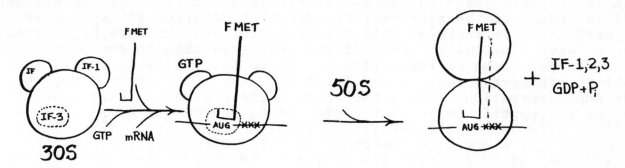

Figure 10.8. Initiation Phase

Messenger RNA binding to the initiation factor-30S subunit complex releases IF3. The 50S subunit then binds to the 30S subunit, mRNA, fmet tRNA, IF1, IF2, GTP complex. IFI and IF2 are released at this stage upon GTP hydrolysis. The fmet-tRNA is positioned correctly (anticodon bound to initiator AUG in mRNA) in the P (peptidyl tRNA) site of the 70S ribosome, poised for the elongation phase of protein synthesis.

Chain elongation involves the binding of a charged (aminoacylated) tRNA to the A (aminoacyl tRNA) site of the ribosome (Figure 10.9).

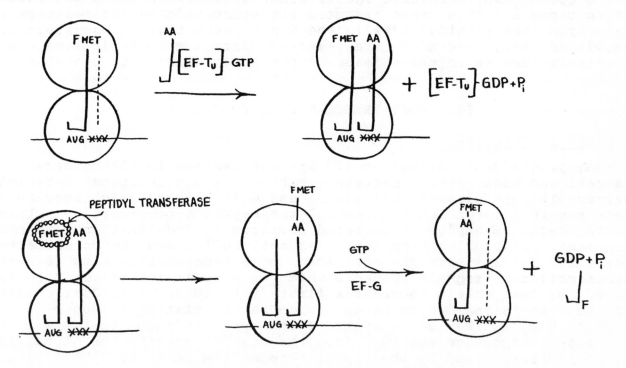

Figure 10.9. Elongation Phase

Aminoacyl tRNA binding depends upon the matching of its anticodon with the triplet codon of the mRNA positioned in the A site. Charged tRNA is complexed with elongation factor **Tu** and **GTP**. Once the charged tRNA is bound to the ribosome, the GTP is hydrolyzed to GDP and Tu is released. Tu is recycled with GTP by factor **Ts**. Peptide bond formation between the amino group of the aminoacyl tRNA in the A site and the carboxyl group of the peptidyl tRNA in the P site is catalyzed by peptidyl transferase, an enzyme located on the 50S subunit. Once peptide bond formation occurs, the tRNA in the A-site now containing the growing polypeptide chain is transferred to the P-site along with concomitant release of the deacylated tRNA in the P-site. This reaction is promoted by **elongation factor G** and GTP. GTP hydrolysis results in the release of G factor. The A site is now vacant and ready for another elongation round. In this manner, the polypeptide chain grows from N-terminal to C-terminal.

The error frequency of protein synthesis (incorporation of an amino acid from an aminoacyl tRNA that was "misread" by the ribosome) is extremely low. This is due to proofreading of the peptidyl-tRNA bound to the P-site in the ribosome. The ribosome checks the match of the codon-anticodon interaction in the P-site and, if not correct, ejects the peptidyl tRNA from the ribosome in a GTP-dependent process. The antibiotic **streptomycin** interferes with this proofreading process and causes misreading of the genetic code.

Polypeptide chain termination occurs when the next available codon in the A site is a termination codon i.e. UAA, UAG, or UGA. A **release factor (RF1** or **RF2)** and GTP binds to the A site containing the terminating codon (which release factor binds is dependent upon the terminating codon). This event triggers the hydrolysis of the polypeptide chain from the peptidyl tRNA in the P site such that the polypeptide, deacylated tRNA, and mRNA are released from the 70 S ribosome. GTP hydrolysis results in the release of the termination factor.

III. POST-TRANSLATIONAL PROCESSING

A. Protein Secretion

Many proteins in animal cells are synthesized in the cytoplasm of the cell and then enter a secretory pathway (rough endoplasmic reticulum, smooth endoplasmic reticulum, and Golgi stack) which results in their eventual insertion into the plasma membrane or secretion outside of the cell. Examples of secreted proteins include insulin, antibodies, many polypeptide hormones, immunoglobulins and serum proteins. Secretory proteins are characterized by a hydrophobic amino acid sequence called a **"signal" sequence** which usually is located at the amino end of the protein (about the first 20 to 30 amino acids). According to the "signal hypothesis", when a translating ribosome on a polysome has synthesized enough of the protein such that its amino acid signal sequence emerges from the large subunit, the signal sequence is recognized by the **signal recognition particle (SRP)** (Figure 10.10).

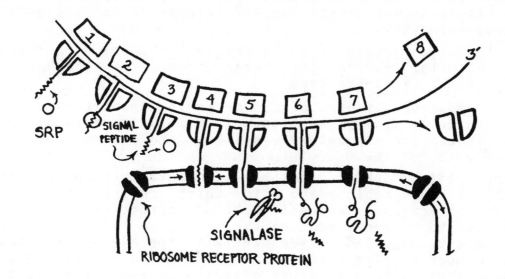

Figure 10.10. Secretion of Proteins through the Membranes of the RER.

The SRP is a ribonucleoprotein containing a 7S RNA and 6-8 specific proteins. The SRP binds to the signal sequence and also to the translating ribosome which results in a halt of protein synthesis. The SRP-ribosome complex is then recognized by a **ribosome receptor protein** or **docking protein** located on the surface of the rough endoplasmic reticulum. This docking protein binds the SRP-ribosome complex to the membrane resulting in the release of SRP and the resumption of translation. Translation causes the amino terminal signal sequence to be inserted through the lipid bilayer into the lumen of the rough ER. The protein is transported inside the rough ER by the translation process. During this process, the signal sequence is usually cleaved off.

B. Glycosylation

Once inside of the lumen of the rough ER, the protein is usually **glycosylated** at specific **asparagine** residues.

The large majority of proteins synthesized in eucaryotic cells undergo glycosylation after the polypeptide backbone has been synthesized. The oligosaccharide side chains of these glycoproteins are formed in two stages.

1. In the rough endoplasmic reticulum (RER) a "mannose core" is attached to certain asparagine residues after the polypeptide chain has been inserted into the lumen of the RER. The mannose core is assembled in the cytosol on an isoprenoid lipid called **dolichol phosphate**. The core is transferred from the dolichol phosphate, through the membrane of the RER, to the asparagine residue. This process can be inhibited by **tunicamycin**.

2. While in the lumen of the RER the mannose core begins to be processed, i.e., certain sugar residues are removed. The partially-

formed glycoprotein is transferred to the *cis*-side of the Golgi apparatus *via* clathrin-coated vesicles and further processing occurs within the Golgi membranes.

3. As the glycoprotein travels through the *medial* and *trans* portions of the Golgi further peripheral sugars are attached until the side chains are in finished form.

4. The completed glycoproteins, depending upon the cell in which they are synthesized, may be transported to lysosomes, secreted from the cell, stored in secretory vesicles or become components of membranes.

5. Glycoproteins destined to become **membrane components** have regions rich in hydrophobic amino acids which are inserted into the lipid bilayer of membranes anchoring the glycoproteins in the membrane. Those destined to be **secreted** from the cell lack these hydrophobic regions and are transported in vesicles with the aid of **microtubules** to the plasma membrane. The vesicles merge with the membrane and release the glycoproteins into the extracellular environment. Glycoproteins that are stored in **secretory granules** are enclosed in vesicular membranes and concentrated for later release upon hormonal stimulation. Certain glycoproteins destined to become lysosomal components are "tagged" during the Golgi transit by attaching phosphate to the 6-position of certain mannose residues. This serves as a signal for their lysosomal destination.

6. The role of the carbohydrate side chains appears to be topological, that is, they offer additional structural components that can act as signals for binding to receptors or to other proteins. They may also control protein folding and the life-time of the glycoprotein.

IV. MODE OF ACTION OF ANTIBIOTICS THAT INHIBIT NUCLEIC ACID METABOLISM AND PROTEIN SYNTHESIS

Actinomycin D: binds to DNA in GC rich regions leading to inhibition of transcription, especially ribosomal RNA.

α-Amanitin: inhibits eucaryotic RNA polymerases I, II, and III to varying degrees.

Chloramphenicol: inhibits peptidyl transferase in bacterial ribosomes and mitochondrial protein synthesis in animal cells.

Cordycepin: inhibits chain elongation during RNA transcription.

Cycloheximide: cause inhibition of the elongation step of protein synthesis on eucaryotic ribosomes.

Diphtheria toxin: covalently modifies elongation factor EF2 in mammalian cells, leading to inhibition of protein synthesis.

Emetine: ipecac alkaloid, inhibits translocation step of protein synthesis in eucaryotic cells.

Erythromycin: inhibits translocation of protein synthesis in bacteria.

Nalidixic acid and **Novobiocin**: inhibit topoisomerases leading to inhibition of DNA synthesis in bacteria.

Puromycin: mimics the aminoacyl region of aminoacyl tRNA and causes premature release of polypeptide chains during protein synthesis in bacterial and eucaryotic cells.

Rifampicin: inhibits bacterial RNA synthesis by binding to RNA polymerase.

Streptomycin: inhibits protein synthesis on bacterial ribosomes; causes misreading of the genetic code.

Tetracycline: blocks the ribosomal A site and inhibits protein synthesis in bacteria.

Tunicamycin: inhibits N-linked glycosylation events in animal cells.

V. REVIEW QUESTIONS ON NUCLEIC ACIDS AND PROTEIN SYNTHESIS

DIRECTIONS: Each of the questions or incomplete statements below is followed by five suggested answers or completions. Select the one that is BEST in each case and fill in the corresponding space on the answer sheet.

1. If GGC is a codon in mRNA (5'-3' direction), which one of the following would be the anticodon (5'-3' direction) in tRNA?

A. GCC
B. CCG
C. CCC
D. CGC
E. GGC

2. Which one of the following antibiotics will inhibit protein synthesis in both procaryotic and eukaryotic organisms?

A. cycloheximide
B. tetracycline
C. streptomycin
D. emetine
E. puromycin

3. A nucleic acid having a sedimentation constant of 4 S and which is capable of being "charged" with an amino acid in a ATP-dependent process is:

A. hnRNA
B. snRNA
C. mRNA
D. tRNA
E. rRNA

4. Which one of the following compounds is used to initiate protein synthesis on mammalian ribosomes?

A. emetine
B. puromycin
C. formylated met-tRNA
D. unformylated met-tRNA
E. unformylated arg-tRNA

5. tRNA molecules have at their 3' termini the sequence:

A. CCA
B. CAA
C. CCC
D. AAC
E. AAA

6. An example of a palindrome in a DNA sequence is:

A. ATGCCG
   TACGGC

B. GGCCGG
   CCGGCC

C. CTAGGG
   GATCCC

D. GAATTC
   CTTAAG

E. TCTGAC
   AGACTG

7. Proteins destined for secretion in mammalian cells contain:

A. A hydrophilic "signal sequence" at their N-terminus.
B. Poly A at their C-terminus
C. A hydrophobic "signal sequence" at their N-terminus.
D. A "cap" structure at their N-terminus
E. Poly A at their N-terminus

8. The sequence of a short duplex DNA is

   5'GAACCTAC3'
   3'CTTGGATG5'

What is the corresponding mRNA sequence transcribed from this region?

A. GAACCTAC
B. CUUGGAUG
C. CAUCCAAG
D. GAACCUAC
E. GUAGGUUC

9. Repair of single-stranded nicks in double stranded DNA and formation of circular molecules from linear DNA pieces can be accomplished by:

A. helicases
B. nucleases
C. restriction enzymes
D. ligases
E. polymerases

10. Which position of the ribose moiety in tRNA is charged with an amino acid?

A. 1'OH
B. 2'OH
C. 3'OH
D. 2'phosphate
E. 3'phosphate

11. Charging of tRNAs with amino acids is performed by which enzyme?

A. ligase
B. phosphatase
C. RNAse
D. peptidase
E. synthetase

12. Amino acids in proteins are covalently linked together by:

A. salt linkages
B. van der Waals interactions
C. peptide bonds
D. hydrophobic bonds
E. phosphodiester bonds

13. The large subunit of mammalian ribosomes has a sedimentation constant of:

A. 40 S
B. 70 S
C. 30 S
D. 80 S
E. 60 S

14. A cluster of ribosomes translating the same mRNA is called a(n):

A. episome
B. monosome
C. polysome
D. spliceosome
E. genome

15. Restriction enzymes:

A. cut single-stranded DNA in a random fashion
B. cut double-stranded DNA in a sequence-specific fashion
C. are used to join two DNA molecules together
D. are not affected by DNA methylation
E. are purified from bacteriophage

16. RNA polymerases require as substrates:

A. ATP, GTP, TTP, CTP
B. ATP, GTP, UTP, CTP
C. dATP, dGTP, dTTP, dCTP
D. an RNA template
E. ribosomes

17. Chain termination codons in mRNA are recognized by which proteins?

A. release factors
B. restriction enzymes
C. elongation factors
D. cap binding protein
E. initiation factors

18. Proteins are synthesized:

A. from the N-terminal to the C-terminal end
B. from the C-terminal to the N-terminal end
C. from rRNA
D. from 19 amino acids
E. in the nucleus

19. A point mutation in a mRNA codon will most likely cause:

A. mRNA degradation
B. inactivation of ribosomes
C. an altered amino acid sequence
D. incomplete transcription
E. inhibition of splicing

20. The DNA consensus sequence "TATA" is a:

A. initiation signal for DNA synthesis
B. initiation signal for RNA synthesis
C. histone binding site
D. ribosome binding site
E. DNA methylation site

21. The nucleosome core contains:

A. 1 copy each of histone H2A, H2B, H3, and H4.
B. 2 copies each of histone H2A, H2B, H3, and H4.
C. DNA covalently linked to histones
D. histone H1
E. RNA polymerase

22. Which one of the following is NOT an example of post-transcriptional modification of mammalian hnRNA?

A. polyadenylation
B. splicing together of exons
C. degradation
D. capping
E. splicing together of introns

23. Which one of the following is not a property of eukaryotic transcription?

A. requires 2 distinct RNA polymerases
B. occurs in the nucleus
C. requires transcription factors
D. occurs in the 5' to 3' direction
E. inhibited by the mushroom poison amanitin

24. Histones are rich in the amino acid(s):

A. cysteine
B. phenylalanine
C. tryptophan
D. leucine and valine
E. lysine and arginine

25. Diphtheria toxin inhibits:

A. protein synthesis initiation
B. protein synthesis elongation
C. protein synthesis termination
D. DNA supercoiling
E. mRNA splicing

26. The signal recognition particle recognizes:

A. RNA polymerase
B. DNA polymerase
C. nucleosomes
D. the N-terminus of secretory proteins
E. poly A

27. Each ribosome in a polysome is:

A. moving in the 3' to 5' direction on the mRNA
B. synthesizing many polypeptide chains
C. synthesizing only one polypeptide chain
D. dissociated
E. inhibited by actinomycin D

28. The AIDS virus (HTLV III) contains an RNA genome. Which enzyme is responsible for converting this genome into DNA in T lymphocytes?

A. Ligase
B. EcoRI
C. DNA polymerase
D. reverse transcriptase
E. RNA polymerase

29. Small nuclear RNAs (snRNA. are involved in:

A. DNA synthesis
B. RNA synthesis
C. splicing
D. aminoacylation
E. recombination

30. The codon UUU is specific for the amino acid:

A. leucine
B. lysine
C. proline
D. phenylalanine
E. hydroxyproline

31. Which nucleic acid is most rapidly degraded in mammalian cells?

A. rRNA
B. hnRNA
C. tRNA
D. rDNA
E. 5S RNA

32. Which nucleic acid is most stable in mammalian cells?

A. mRNA
B. 45S rRNA
C. duplex DNA
D. thymine dimers
E. precursor tRNA

33. Denatured human lymphocyte DNA will not hybridize with human:

A. lymphocyte rRNA
B. kidney tRNA
C. denatured mitochondrial DNA
D. denatured liver DNA
E. brain mRNA

34. Euchromatin is:

A. capable of being transcribed
B. is highly condensed
C. found during mitosis
D. found in metaphase chromosomes
E. is ribosome associated

35. Tissue cells incubated with [³H] thymidine will most rapidly accumulate the radioactive isotope in:

A. mitochondria
B. ribosomes
C. hnRNA
D. rough ER
E. nuclei

36. DNA polymerase utilizes which compounds as substrates?

A. CTP, UTP, GTP, ATP
B. dCTP, dUTP, dGTP, dATP
C. rRNA
D. mRNA
E. dATP, dCTP, dTTP, dGTP

37 The genetic code is:

A. non-degenerate
B. species specific
C. triplet
D. punctuated
E. translated by RNA polymerase

38. DNA contains:

A. pyrimidine ribose monophosphates
B. purine deoxyribose diphosphates
C. purine deoxyribose monophosphates
D. purine deoxyhexose monophosphates
E. 5'-2' phosphodiester bonds

39. A codon mutation resulting from a deletion of a nucleotide base results in a:

A. translocation
B. inversion
C. frameshift change
D. transversion
E. transition

DIRECTIONS: Each set of lettered headings below is followed by a list of numbered words or phrases. On the answer sheet, for each numbered word or phrase fill in space
   A if the item is associated with (A) only
   B if the item is associated with (B) only
   C if the item is associated with both (A) and (B)
   D if the item is associated with neither (A) nor (B)

Questions 40 - 43:

   A. transfer RNA (rRNA)
   B. ribosomal RNA (rRNA)
   C. Both
   D. Neither

40. In bacteria, is found as 16 S and 23 S molecules

41. forms ester bonds with activated amino acids

42. contains amino acid-specific codons

43. synthesis is inhibited by actinomycin D

Questions 44 - 47:

   A. EF-G
   B. peptidyl transferase
   C. Both
   D. Neither

44. forms peptide bond

45. inhibited by chloramphenicol in bacteria

46. requires GTP

47. involved ribosome translocation

DIRECTIONS: For each of the questions or incomplete statements below, ONE or MORE of the answers or completions is correct. On the answer sheet fill in space

A if only 1, 2, and 3 are correct
B if only 1 and 3 are correct
C if only 2 and 4 are correct
D if only 4 is correct
E if all are correct

FILL IN ONLY ONE SPACE ON YOUR ANSWER SHEET FOR EACH QUESTION

Directions Summarized				
(A) 1,2,3 only	(B) 1,3 only	(C) 2,4 only	(D) 4 only	(E) All are correct

48. GTP is NOT used in what steps of protein synthesis?

1. binding of aminoacyl-tRNA to elongation factor Tu
2. association of ribosomal subunits during initiation of protein synthesis
3. elongation of factor G-dependent translocation
4. action of peptidyl transferase

49. Which of the following statements about mammalian mRNA are correct?

1. contains an unusual structure at its 5' end
2. has a long stretch of poly A at its 3'end
3. is transcribed as a hnRNA precursor
4. is transcribed by RNA polymerase I

50. What eukaryotic RNAs are synthesized in the nucleolus?

1. 5 S ribosomal RNA
2. 18 S ribosomal RNA
3. tRNA
4. 28 S ribosomal RNA

51. Cell surface glycoproteins:

1. are responsible for some cellular antigenicity
2. are formed on the cell surface by extracellular enzymes
3. are formed by sequential passage through the rough ER, smooth ER, and Golgi apparatus.
4. are synthesized free in the cytoplasm

52. Alterations in which nucleic acids would be found in sickle cell anemia?

1. hnRNA
2. mRNA
3. DNA
4. rRNA

53. Nucleic acids:

1. are helical
2. absorb more UV light than nucleotides
3. contain more phosphorus than oxygen
4. absorb less UV light than nucleotides

54. DNA ligase:

1. joins segments of DNA during DNA synthesis
2. joins the ends of RNA
3. important enzyme in DNA recombinant technology
4. catalyzes the condensation of a 5'-nucleotide phosphate group with a 2'-sugar OH group

55. Protein synthesis in vitro requires:

1. aminoacylated tRNAs
2. mRNA
3. Mg^{++}
4. dATP

56. The genetic code is:

1. degenerate
2. highly conserved
3. triplet
4. overlapping

FILL IN ONLY ONE SPACE ON YOUR ANSWER SHEET FOR EACH QUESTION

Directions Summarized				
(A) 1,2,3 only	(B) 1,3 only	(C) 2,4 only	(D) 4 only	(E) All are correct

57. When DNA is subjected to high pH, properties which are changed include:

1. viscosity
2. helicity
3. absorption of visible light
4. sequence

58. Thymine dimers:

1. are removed enzymatically in the cell
2. are formed by UV irradiation
3. are not removed in patients with xeroderma pigmentosum
4. form bridges between 2 complementary DNA strands

59. A DNA template is utilized in which process?

1. transcription
2. translation
3. semi-conservative replication
4. aminoacylation

60. Which of the following proteins require post translational modification for proper functioning?

1. insulin
2. immunoglobulins
3. thrombin
4. pepsin

61. A nucleic acid was found to contain upon analysis the following base composition:

   A=16%; G=39%; C=22%; and T=23%

This nucleic acid is most likely:

1. double stranded DNA
2. tRNA
3. single stranded RNA
4. single stranded DNA

62. tRNA molecules have structural elements which recognize:

1. mRNA codons
2. aminoacyl synthetases
3. elongation factor Tu
4. restriction enzymes

63. An increase in the DNA mutation rate of an organism can result from:

1. mutant RNA polymerase
2. mutant DNA polymerase
3. too many polysomes
4. carcinogens

64. Reverse transcriptase will synthesize DNA from what templates?

1. HTLV genome
2. double stranded DNA
3. single stranded RNA
4. double stranded RNA

65. Histones:

1. main protein component of nucleosomes
2. are not very conserved
3. can be phosphorylated
4. bind mRNA

66 DNA synthesis in bacteria and higher organisms:

1. requires RNA synthesis
2. inhibited by cycloheximide
3. releases pyrophosphate
4. is unidirectional

67. Transfer RNA (tRNA):

1. has a 5' terminus which is aminoacylated
2. is 120 nucleotides in length
3. has 1-2% unusual bases
4. hybridizes to mRNA

VI. ANSWERS TO QUESTIONS ON NUCLEIC ACIDS AND PROTEIN SYNTHESIS

1. A	26. D	51. B
2. E	27. C	52. A
3. D	28. D	53. A
4. D	29. C	54. B
5. A	30. D	55. A
6. D	31. B	56. A
7. C	32. C	57. A
8. D	33. C	58. A
9. D	34. A	59. B
10. C	35. E	60. E
11. E	36. E	61. D
12. C	37. C	62. A
13. E	38. C	63. C
14. C	39. C	64. B
15. B	40. B	65. B
16. B	41. A	66. B
17. A	42. D	67. D
18. A	43. C	
19. C	44. B	
20. B	45. B	
21. B	46. A	
22. E	47. A	
23. A	48. D	
24. E	49. A	
25. B	50. C	

11. HUMAN GENETICS: INBORN ERRORS OF METABOLISM

Wai-Yee Chan

I. CARBOHYDRATE METABOLISM

A. Diabetes Mellitus

Molecular defect: Insulin deficiency.
Pathway affected: Inappropriate gluconeogenesis.
Diagnosis: Hyperglycemia, glucosuria and ketogenesis.
Genetics: Multifactorial.
Treatment: (1)* Insulin.

B. Fructose Metabolism

1. **Hereditary fructose intolerance.**
Molecular defect: Deficient fructose-1-phosphate aldolase b.
Pathway affected: $F-1-P^* \longrightarrow DHA-P + \alpha-Glycero-P$.
Diagnosis: Hypoglycemia.
Genetics: Autosomal recessive.
Treatment: (4) Fructose and sucrose.

2. **Fructose-1,6-diphosphatase deficiency.**
Molecular defect: Deficient F-1,6-diPase.
Pathway affected: $F-1,6-diP \longrightarrow F-6-P + P_i$
Diagnosis: Hypoglycemia and hyperlacticacidemia.
Genetics: Autosomal recessive.
Treatment: (4) Fructose.

C. Galactose Metabolism

1. **Galactokinase deficiency**
Molecular defect: Deficient galactokinase.
Pathway affected: $Gal \longrightarrow Gal-1-P$.
Diagnosis: Galactose and galactitol in urine.
Genetics: Autosomal recessive.

2. **Classical galactosemia**
Molecular defect: Deficient Gal-1-P uridyltransferase.
Pathway affected: $UDPGlc + Gal-1-P \longrightarrow UDPGal + G-1-P$
Diagnosis: Accumulation of galactose, galactitol and Gal-1-P.
Genetics: Autosomal recessive.
Treatment: (4) Galactose.

D. Glycogen Storage Diseases

1. **Von Gierke**
Molecular defect: Deficient glucose-6-phosphatase.
Pathway affected: $G-6-P \longrightarrow Glc + P_i$

* Refer to categories under XIV. Treatment of Inherited Metabolic Diseases, page 190.
* See XV. List of Abbreviations, page 191.

Diagnosis: Hypoglycemia, increased lactic acid and hyperlipidemia; increased glycogen with normal structure.
Genetics: Autosomal recessive.

2. **Pompe**
   Molecular defect: Deficient lysosomal Acid α 1,4-Glucosidase (Acid maltase).
   Pathway affected: Maltose, linear oligosaccharides, glycogen $\longrightarrow$ glucose.
   Diagnosis: Increased glycogen of normal structure.
   Genetics: Autosomal recessive.

3. **Cori**
   Molecular defect: Deficient debranching enzyme (amylo-1,6-glucosidase).
   Pathway affected: Limit dextrin $\longrightarrow$ glucose.
   Diagnosis: Glycogen with shorter outer chains and increased branching.
   Genetics: Autosomal recessive.

4. **Andersen**
   Molecular defect: Deficient branching enzyme (α-1,4-glucan: α-1,4-glucan 6-glucosyl-transferase.
   Genetics: Autosomal recessive.

5. **McArdle**
   Molecular defect: Deficient muscle phosphorylase.
   Pathway affected: Glycogen $\longrightarrow$ G-1-P.
   Diagnosis: Increased glycogen (M) of normal structure.
   Genetics: Autosomal recessive.

6. **Hers**
   Molecular defect: Deficient liver phosphorylase.
   Pathway affected: Glycogen $\longrightarrow$ G-1-P.
   Genetics: Autosomal recessive.

7. **Tarui**
   Molecular defect: Deficient muscle phosphofructokinase.
   Pathway affected: F-6-P $\longrightarrow$ F-1,6-diP
   Diagnosis: Increased glycogen of normal structure.
   Genetics: Autosomal recessive.

8. **Hepatic phosphorylase b kinase deficiency**
   Molecular defect: Deficient liver phosphorylase b kinase.
   Pathway affected: phosphorylase b $\longrightarrow$ phosphorylase a.
   Diagnosis: Increased glycogen.
   Genetics: X-linked.

E. Others

1. **Pyruvate carboxylase deficiency**
   Molecular defect: Deficient pyruvate carboxylase.
   Pathway affected: pyruvate + CO_2 $\longrightarrow$ oxaloacetate.
   Diagnosis: Increased lactate.
   Genetics: Autosomal recessive.

2. **Congenital lactase deficiency**
   Molecular defect: Deficient lactase.
   Pathway affected: Lactose $\longrightarrow$ galactose + glucose.
   Diagnosis: Watery diarrhea, glucosuria, lactosuria.
   Genetics: Autosomal recessive.
   Treatment: (4) Lactose.

II. AMINO ACID METABOLISM

A. <u>Disorders of Sulfur Amino Acid Metabolism</u>

   1. **Homocystinuria**
      Molecular defect: Deficient cystathionine-b-synthase.
      Pathway affected: Homocystine + Ser $\longrightarrow$ cystathionine
      Diagnosis: Increased methionine (blood) and homocystine (urine).
      Genetics: Autosomal recessive.
      Treatment: (4) methionine.

   2. **Combined deficiency of 5-methyltetrahydrofolate : homocysteine methyltransferase, and of methylmalonyl-CoA mutase**
      Molecular defect: Deficient activity of enzymes.
      Pathway affected: Impaired synthesis of AdoCbl and MeCbl.
      Diagnosis: Homocystinuria and methylmalonic acidemia
      Genetics: Autosomal recessive.

B. <u>Disorders of Aromatic Amino Acids</u>

   1. **Phenylketonuria**
      Molecular defect: Deficient phenylalanine hydroxylase.
      Pathway affected: Phe $\longrightarrow$ Tyr.
      Diagnosis: Increased phenylalanine (blood) and phenylpyruvate (urine).
      Genetics: Autosomal recessive.
      Treatment: (4) Phenylalanine.

   2. **Richner-Hanhart**
      Molecular defect: Deficient hepatic tyrosine aminotransferase.
      Pathway affected: Tyr $\longrightarrow$ p-hydroxyphenylpyruvic acid
      Diagnosis: Tyrosinemia, tyrosinuria and increased urinary phenolic acids.
      Genetics: Autosomal recessive.
      Treatment: (4) tyrosine and phenylalanine.

   3. **Alkaptonuria**
      Molecular defect: Deficient homogentisic acid oxidase.
      Pathway affected: Homogentisic acid $\longrightarrow$ acetoacetic acid + fumaric acid.
      Genetics: Autosomal recessive.

   4. **Albinism**
      Molecular defect: Deficient tyrosinase.
      Pathway affected: Defect in melanin synthesis.
      Diagnosis: No detectable melanin.
      Genetics: Autosomal recessive.

C. <u>Disorders of the Urea Cycle</u>

1. **Carbamyl phosphate synthetase deficiency**
   Molecular defect: Deficient enzyme activity.
   Pathway affected: $NH_3 + HCO_3^- + ATP \longrightarrow$ carbamoyl P + ADP.
   Diagnosis: Hyperammonemia.
   Genetics: Autosomal recessive.

2. **Ornithine transcarbamylase deficiency**
   Molecular defect: Deficient enzyme activity.
   Pathway affected: Ornithine + CP $\longrightarrow$ citrulline.
   Diagnosis: Hyperammonemia and orotic aciduria.
   Genetics: X-liked dominant.

3. **Citrullinemia**
   Molecular defect: Deficient argininosuccinate synthetase.
   Pathway affected: Asp + citrulline $\longrightarrow$ Argininosuccinate.
   Diagnosis: Citrullinemia.
   Genetics: Autosomal recessive.

4. **Argininosuccinic aciduria**
   Molecular defect: Deficient argininosuccinase.
   Pathway affected: Argininosuccinate $\longrightarrow$ Arg + Fumarate
   Diagnosis: Argininosuccinic aciduria.
   Genetics: Autosomal recessive.

5. **Argininemia**
   Molecular defect: Deficient arginase.
   Pathway affected: Arg $\longrightarrow$ urea + ornithine.
   Diagnosis: Hyperargininemia.
   Genetics: Autosomal recessive.
   Treatment: (4) Arginine.

D. <u>Disorder of Histidine Metabolism: Histidinemia</u>
   Molecular defect: Deficient histidase.
   Pathway affected: His $\longrightarrow$ urocanic acid.
   Diagnosis: histidinemia and histidinuria.
   Genetics: Autosomal recessive.

E. <u>Disorders of Branched-chain Amino Acids</u>

1. **Maple syrup urine disease**
   Molecular defect: Deficient branched chain α-ketoacid decarboxy-
       lase.
   Pathway affected: Oxidative decarboxylation of the branched
       chain keto acids.
   Diagnosis: Increased Ile, Leu and Val (blood and urine).
   Genetics: Autosomal recessive.
   Treatment: (4) Ile, Leu and Val.

2. **Isovaleric acidemia**
   Molecular defect: Deficient isovaleryl CoA dehydrogenase.
   Pathway affected: isovaleric acid $\longrightarrow$ 3-methylcrotonic acid.
   Diagnosis: Isovaleric acidemia and excessive urinary isovaleryl-
       glycine.

Genetics: Autosomal recessive.
Treatment: (4) Leucine.

3. **3-Hydroxy-3-methylglutaryl-CoA (HMG-CoA) lyase deficiency**
   Molecular defect: Deficient HMG-CoA lyase.
   Pathway affected: HMG-CoA $\longrightarrow$ acetyl CoA + acetoacetic acid.
   Diagnosis: Hypoglycemia and metabolic acidosis.
   Genetics: Autosomal recessive.

F. Genetic Organic Acidemias

   1. **Propionic acidemia**
      Molecular defect: Deficient propionyl-CoA carboxylase.
      Pathway affected: Propionyl-CoA $\longrightarrow$ D-methylmalonyl CoA
      Diagnosis: Propionic acid (blood and urine), ketotic hyper-
         glycinemia and ketoacidosis.
      Genetics: Autosomal recessive.

   2. **Cobalamin-unresponsive methylmalonic acidemia**
      Molecular defect: Mutated methylmalonyl CoA mutase;
         2 subgroups: mut- and mut$^{\circ}$
      Diagnosis: methylmalonic acid (MMA) (blood and urine).
      Genetics: Autosomal recessive.
      Treatment: (4) Protein.

   3. **Cobalamin-responsive methylmalonic acidemia**
      Molecular defect: Deficient ATP : Cob(I)alamin adenosyl trans-
         ferase in CblB
            4 subgroups: CblA, B, C, and D.
      Pathway affected: OH-Cbl $\longrightarrow$ AdoCbl in CblA and CblB
                        OH-Cbl $\longrightarrow$ AdoCbl and MeCbl in CblC and CblD
      Genetics: Autosomal recessive.
      Treatment: (3) OH-Cbl.

III. LIPOPROTEIN AND LIPID METABOLISM

1. **Familial lipoprotein lipase deficiency**
   Molecular defect: Deficient lipoprotein lipase
   Pathway affected: Hydrolysis of glyceride ester bonds.
   Diagnosis: Accumulation of chylomicron triglyceride.
   Genetics: Autosomal recessive.
   Treatment: (4) Lipid.

2. **Familial hypercholesterolemia**
   Molecular defect: Abnormal LDL receptors; 3 mutations:
                     a. no receptor binding of LDL
                     b. reduced receptor binding of LDL
                     c. no internalization of bound LDL
   Pathway affected: Cellular uptake and delivery of LDL to lyso-
      some.
   Diagnosis: High cholesterol in plasma.
   Genetics: Autosomal dominant.
   Treatment: (4) Dietary cholesterol.

3. **Familial lecithin:cholesterol acyltransferase deficiency (LCAT)**
   Molecular defect: Deficient LCAT.
   Pathway affected: Cholesterol $\longrightarrow$ cholesteryl ester.
   Diagnosis: Abnormal plasma lipoprotein pattern.
   Genetics: Autosomal recessive.

4. **Abetalipoproteinemia and familial hypobetalipoproteinemia**
   Molecular defect: Abnormal synthesis or intracellular assembly
       of ApoB with lipid.
   Pathway affected: Transport of triglyceride.
   Diagnosis: Absence of chylomicrons, VLDL and LDL.
   Genetics: Autosomal recessive.

5. **Tangier**
   Molecular defect: Defective regulation of HDL synthesis or cata-
       bolism (LP containing APO A-I and Apo A-II)
   Pathway affected: Transport of cholesterol.
   Diagnosis: Accumulation of cholesteryl esters in tissue; defi-
       cient plasma HDL.
   Genetics: Autosomal recessive.

6. **Familial dysbetalipoproteinemia**
   Molecular defect: Deficient Apo E-3
   Pathway affected: Uptake of remnant lipoproteins.
   Diagnosis: Abnormal Apo E pattern of plasma VLDL.
   Genetics: Autosomal recessive.

IV. DISORDERS OF LYSOSOMAL ENZYMES

A. <u>The Mucopolysaccharidoses</u>

1. **Hurler, MPS 1H**
   Molecular defect: Deficient α-L-iduronidase.
   Pathway affected: Degradation of dermatan sulfate and heparin
       sulfate
   Genetics: Autosomal recessive.
   Treatment: (6) bone marrow transplant
               (1) plasma infusion, enzyme replacement

2. **Scheie, MPS 1S (Formerly MPS V)**
   Molecular defect: Deficient α-L-iduronidase.
   Pathway affected: Degradation of dermatan sulfate and heparin
       sulfate
   Genetics: Autosomal recessive.

3. **Hunter, MPS II**
   Molecular defect: Deficient iduronate sulfatase
   Pathway affected: Sulfated iduronate $\longrightarrow$ iduronate
   Genetics: X-linked recessive.
   Treatment: (1) plasma infusion, enzyme replacement.

4. **Sanfilippo A, MPS IIIA**
   Molecular defect: Deficient heparan N-sulfatase.

Genetics: Autosomal recessive.
Treatment: (1) plasma infusion, enzyme replacement.

5. **Sanfilippo B**, MPS III A
   Molecular defect: Deficient N-acetyl-α-D-glucosaminidase.
   Diagnosis: Urinary heparan sulfate
   Genetics: Autosomal recessive.
   Treatment: (1) plasma infusion, enzyme replacement.

6. **Sanfilippo C**, MPS III C
   Molecular defect: Deficient acetyl CoA:α-glucosaminide N-acetyl-
      transferase.
   Pathway affected: Acetylation of glucosamine.
   Diagnosis: Urinary heparan sulfate.
   Genetics: Autosomal recessive.

7. **Sanfilippo D**, MPS IIID
   Molecular defect: Deficient N-acetyl-α-D-glucosaminide-6-sulfa-
      tase.
   Pathway affected: Release of sulfate from N-acetyl-glucosamine
   of heparan sulfate.
   Diagnosis: Urinary heparan sulfate.
   Genetics: Autosomal recessive.

8. **Morquio A**, MPS IV A
   Molecular defect: Deficient galactosamine-6-sulfate sulfatase.
   Pathway affected: Desulfation of galactose-6-sulfate.
   Diagnosis: Urinary keratan sulfate.
   Genetics: Autosomal recessive.

9. **Morquio B**, MPS IV B
   Molecular defect: Deficient ß-galactosidase.
   Pathway affected: Hydrolysis of ß-galactosyl residues of keratan
      sulfate.
   Diagnosis: Urinary keratan sulfate.
   Genetics: Autosomal recessive.

10. **Maroteaux-Lamy**, MPS VI
    Molecular defect: Deficient Arylsulfatase B.
    Pathway affected: Desulfation of N-acetylgalactosamine-4-sul-
       fate.
    Diagnosis: Urinary dermatan sulfate.
    Genetics: Autosomal recessive.

11. **Sly**, MPS VII
    Molecular defect: Deficient ß-glucuronidase.
    Pathway affected: Catabolism of polymers containing ß-linked
       glucuronic acid residues.
    Diagnosis: Urinary dermatan sulfate and heparan sulfate.
    Genetics: Autosomal recessive.

B. <u>Oligosaccharidoses</u>

   1. **I-cell disease**, mucolipidosis II, ML II, and **pseudo-Hurler poly-
      dystrophy**, ML III
      Molecular defect: Deficient phosphorylation of mannose residues
         of lysosomal hydrolases.

Pathway affected: Receptor-mediated uptake of lysosomal hydro-
   lases.
Diagnosis: Elevated activities of lysosomal enzymes in serum.
Genetics: Autosomal recessive.

2. **Sialidosis, ML I**
   Molecular defect: Deficient neuraminidase.
   Pathway affected: Cleavage of 2,3- and 2,6-linked N-acetylneura-
      minic acid.
   Diagnosis: Urinary oligosaccharides.
   Genetics: Autosomal recessive.

3. **Mannosidosis**
   Molecular defect: Deficient acidic α-mannosidase.
   Pathway affected: Man (α-1,3, α-1,6) $\longrightarrow$ Man
   Diagnosis: Reduced enzyme activity in serum.
   Genetics: Autosomal recessive.

4. **Fucosidosis**
   Molecular defect: Deficient α-L-fucosidase.
   Pathway affected: Fuc (α-1,6) $\longrightarrow$ GlcNAc
   Diagnosis: Reduced enzyme activity in leucocytes.
   Genetics: Autosomal recessive.

5. **Aspartylglycosaminuria**
   Molecular defect: Deficient Aspartylglycosaminidase.
   Pathway affected: β-Aspartylglycosylamine $\longrightarrow$ N-acetylgluco-
      samine + NH_3
   Diagnosis: Urinary aspartylglucosamine.
   Genetics: Autosomal recessive.

C. <u>Sphingolipidoses (The Gangliosidoses)</u>

1. **G_{M1}-gangliosidosis types 1, 2 and 3**
   Molecular defect: Mutated β-gangliosidase A.
   Pathway affected: Ganglioside $G_{M1} \longrightarrow G_{M2}$
   Diagnosis: Reduced enzyme activity in leucocytes.
   Genetics: Autosomal recessive.

2. **Tay-Sachs, G_{M2}-gangliosidosis type 1**
   Molecular defect: Deficient hexosaminidase A; mutated α chain.
   Pathway affected: $G_{M2} \longrightarrow$ N-acetyl-β-D-galactosamine.
   Diagnosis: Deficient Hex A in serum and leucocytes.
   Genetics: Autosomal recessive.

3. **Sandhoff, G_{M2}-gangliosidosis type 2**
   Molecular defect: Deficient Hex A and B; mutated β chain.
   Pathway affected: $G_{M2} \longrightarrow$ N-acetyl-β-D-galactosamine.
   Diagnosis: Deficient activities of Hex A and Hex B in serum.
   Genetics: Autosomal recessive.

D. <u>Other Sphingolipidoses</u>

1. **Gaucher, types 1, 2 and 3**
   Molecular defect: Deficient β-glucocerebrosidase
   Pathway affected: Glucocerebroside $\longrightarrow$ Glc

Diagnosis: Reduced enzyme activity in leucocytes
Genetics: Autosomal recessive (all three types).

2. **Krabbe, globoid cell leucodystrophy**
Molecular defect: Deficient galactocerebroside ß-galactosidase.
Pathway affected: galactocerebroside ⟶ ceramide + Gal.
Diagnosis: deficient enzyme activity in leucocytes.
Genetics: Autosomal recessive.

3. **Niemann-Pick**, types A and B
Molecular defect: Deficient sphingomyelinase.
Pathway affected: Sphingomyelin ⟶
                  ceramide + phosphorylcholine.
Diagnosis: Deficient enzyme activity in leucocytes
Genetics: Autosomal recessive.

4. **Metachromatic leukodystrophy**
Molecular defect: Deficient arylsulfatase A
Pathway affected: Catabolism of sulfatides and sulfogalactoglyc-
    erolipids.
Diagnosis: Deficient enzyme activity in leucocytes
Genetics: Autosomal recessive.

5. **Multiple sulfatase deficiency**
Molecular defect: Mutated regulatory gene affecting expression
    of 9 distinct enzymes including ASA, ASB, ASC, cholesteryl
    sulfatase, dehydroepiandrosterone sulfatase, iduronide-2-sul-
    fate sulfatase, heparan-N-sulfamidase, N-acetylgalactosamine-
    4-sulfate sulfatase, and N-acetylglucosamine-6-sulfate sulfa-
    tase.
Pathway affected: Metabolism of sulfate-containing glycolipids,
    mucopolysaccharides, and steroids.
Diagnosis: Increased sulfatide in urine.
Genetics: Autosomal recessive.

6. **Fabry**
Molecular defect: Deficient α-galactosidase A.
Pathway affected: Gal-Gal-Glc-Cer ⟶ Gal-Glc-Cer + Gal.
Diagnosis: Increased tissue Gal-Gal-Glc-Cer; deficient α-galac-
    tosidase A (plasma).
Genetics: X-linked.

7. **Farber**
Molecular defect: Deficient acid ceramidase.
Pathway affected: Ceramide ⟶ sphingosine + fatty acid.
Diagnosis: accumulation of ceramide in tissue.
Genetics: Autosomal recessive.

8. **Wolman, and cholesteryl ester storage disease**
Molecular defect: Deficient acid lipase.
Pathway affected: Hydrolysis of cholesteryl esters.
Diagnosis: Accumulation of cholesteryl esters and triglyceride
    in tissues.
Genetics: Autosomal recessive.

V. PURINE AND PYRIMIDINE METABOLISM

1. **Lesch-Nyhan**
   Molecular defect: Deficient hypoxanthine-guanine phosphoribosyl-
      transferase (HGPRT).
   Pathway affected: PRPP + hypoxanthine $\longrightarrow$ IMP.
                     PRPP + guanine $\longrightarrow$ GMP.
   Diagnosis: Increased uric acid (blood and urine).
   Genetics: X-linked.

2. **Adenine phosphoribosyltransferase (APRT) deficiency**
   Molecular defect: Mutated structural gene of APRT.
   Pathway affected: Adenine $\longrightarrow$ AMP.
   Diagnosis: Increased urinary adenine, 2,8-DHA and 8-HA.
   Genetics: Autosomal recessive.

3. **Adenosine deaminase (ADA) deficiency**
   Molecular defect: Mutated structural gene of ADA.
   Pathway affected: Adenosine $\longrightarrow$ inosine.
   Diagnosis: Combined immunodeficiency disease (CID).
   Genetics: Autosomal recessive.
   Treatment: (6) bone marrow transplant.
              (1) administer erythrocytes that have missing enzyme.

4. **Purine nucleoside phosphorylase (PNP) deficiency**
   Molecular defect: Mutated structural gene of PNP.
   Pathway affected: Hypoxanthine $\longrightarrow$ deoxyinosine.
                     Guanine $\longrightarrow$ guanosine
   Diagnosis: Deficient cell-mediated immunity.
   Genetics: Autosomal recessive.

5. **Hereditary xanthinuria**
   Molecular defect: Deficient xanthine oxidase.
   Pathway affected: Hypoxanthine $\longrightarrow$ xanthine $\longrightarrow$ uric acid.
   Diagnosis: Decreased uric acid (blood and urine).
              Increased xanthine and hypoxanthine (urine).
   Genetics: Autosomal recessive.

6. **Hereditary orotic aciduria,** Type I
   Molecular defect: Deficient orotate phosphoribosyl transferase
      (OPRT) and orotidine 5'-phosphate decarboxylase (ODC), proba-
      bly due to defective single multifunctional protein or one
      abnormal protein affecting the total complex.
   Pathway affected: PRPP $\longrightarrow$ orotidine 5'-phosphate (OPRT);
                     orotidine 5'-phosphate $\longrightarrow$ uridine 5'-P (ODC).
   Diagnosis: orotic aciduria.
   Genetics: Autosomal recessive.

7. **Xeroderma pigmentosum**
   Molecular defect: Excision repair of pyrimidine dimers.
   Pathway affected: Incision of UV-damaged (pyrimidine dimer for-
      mation) DNA (Groups A-G), post-replication repair (Variant).
   Genetics: Autosomal recessive.

VI. STEROID METABOLISM

A. Female Pseudohermaphroditism

1. **21-Hydroxylase deficiency**
   Molecular defect: Deficient 21-Hydroxylase.
   Pathway affected: C-21 hydroxylation.
   Diagnosis: Increased 17-ketosteroids and 21-deoxysteroids (blood
       and urine).
   Genetics: Autosomal recessive.

2. Others

Deficient Enzyme	Compound	Genetics
11β-Hydroxylase	17-Ketosteroids (urine)	Autosomal recessive.
3β-HSD	DHEA (blood)	Autosomal recessive.

B. Male Pseudohermaphroditism, Testosterone Deficiency

1. **5α-Reductase deficiency**
   Molecular defect: Deficient enzyme from more than one mutation.
   Pathway affected: Testosterone $\longrightarrow$ DHT.
   Diagnosis: Elevated ratio of testosterone/DHT.
   Genetics: Autosomal recessive.

2. Other enzyme defects

Deficient Enzyme	Compound	Genetics
20,22-Desmolase	All steroids	Autosomal recessive
17-Hydroxylase	Testosterone low	Autosomal recessive
17,20-Desmolase	Pregnanetriolone	
17-Dehydrogenase	Δ^4-androstenedione/ testosterone ratio	

3. Androgen receptor defects
   **Complete and incomplete testicular feminization, Reifenstein
       syndrome and the infertile male syndrome.**
   Molecular defect: Abnormal or absent androgen receptor protein.
   Pathway affected: Binding of androgens to receptors.
   Diagnosis: Absent binding of DHT to receptors.
   Genetics: X-linked recessive.

VII. METAL METABOLISM

A. Copper Metabolism

1. **Wilson**
   Molecular defect: Biliary excretion and ceruloplasmin incorpora-
       tion of copper, copper toxicity.
   Pathway affected: Transport of copper from liver to other tis-
       sues and cuproenzymes.
   Diagnosis: Liver copper, increased urinary copper.
   Genetics: Autosomal recessive.
   Treatment: (2) penicillamine
              (6) liver transplant.

2. **Menkes**
   Molecular defect: not defined yet, copper deficiency.
   Pathway affected: Intestinal absorption and tissue utilization
      of copper.
   Diagnosis: Low serum copper.
   Genetics: X-linked.

B. Iron: **Idiopathic hemochromatosis**

   Molecular defect: not yet defined, iron toxicity.
   Pathway affected: Absorption of iron.
   Diagnosis: Increased plasma iron and ferritin.
   Genetics: Autosomal recessive.
   Treatment: (2) Phlebotomy.

C. Zinc: **Acrodermatitis enteropathica**

   Molecular defect: not yet defined, zinc deficiency.
   Pathway affected: Absorption of zinc.
   Diagnosis: Decreased serum zinc.
   Genetics: Autosomal recessive.

VIII. PORPHYRIN AND HEME METABOLISM

A. The Porphyrias

   1. **Erythropoietic protoporphyria** (EPP)
      Molecular defect: Deficient ferrochelatase.
      Pathway affected: Last step of heme synthesis.
      Diagnosis: Increased protoporphyrin (plasma and feces).
      Genetics: Autosomal dominant.

   2. **Acute intermittent porphyria** (AIP)
      Molecular defect: Deficient porphobilinogen deaminase.
      Pathway affected: porphobilinogen $\longrightarrow$ hydroxymethylbilane.
      Diagnosis: Increased δ-aminolevulinic acid.
      Genetics: Autosomal dominant.

   3. **Hereditary coproporphyria** (HCP)
      Molecular defect: Deficient coproporphyrinogen III oxidase.
      Pathway affected: coproporphyrinogen $\longrightarrow$ protoporphyrinogen.
      Diagnosis: Urinary δ-aminolevulinic acid and porphobilinogen.
      Genetics: Autosomal dominant.

   4. **Porphyria cutanea tarda** (PCT)
      Molecular defect: Deficient hepatic uroporphyrinogen decarboxy-
         lase.
      Pathway affected: uroporphyrinogen $\longrightarrow$ coproporphyrinogen.
      Diagnosis: Increased urinary isocoproporphyrin series.
      Genetics: Autosomal dominant.

B. Bilirubin Metabolism: **Hyperbilirubinemia, Crigler-Najjar Syndrome**,
   Type 1
      Molecular defect: absent hepatic bilirubin:UDP-glucuronyl trans-
         ferase.

Pathway affected: Chloral hydrate + trichloroethanol + salicy-
    late ⟶ glucuronide.
Diagnosis: Increased unconjugated bilirubin.
Genetics: Autosomal recessive.
Treatment: (2) Exchange transfusion.

C. Acatalasemia
Molecular defect: Mutated catalase.
Pathway affected: H donor + H_2O_2 ⟶ oxidized donor + H_2O.
Diagnosis: Deficient catalase activity in blood.
Genetics: Autosomal recessive.

IX. HEMOGLOBINOPATHIES

A. Hereditary Methemoglobinemia

1. **NADH cytochrome b₅ reductase deficiency.**
Molecular defect: Deficient NADH dehydrogenase; deficiency of
    reducing enzyme for converting met-Hb back to normal Hb.
Pathway affected: NADH + e^- ⟶ cytochrome b_5 + e^- ⟶ met-Hb.
Diagnosis: Reduced enzyme activity in leucocytes.
Genetics: Autosomal recessive.

2. **M Hemoglobin defect**
Molecular defect: Single amino acid substitutions in hemoglobin
    molecule in the immediate environment of heme iron. Collec-
    tively known as hemoglobins M (Hb M), these variants stabi-
    lize the heme group and hold iron in the ferric state. Five
    variants with substitutions in α or β chain.
Pathway affected: Oxygen binding.
Diagnosis: Tissue anoxia and cyanosis; Hb M cannot accept oxy-
    gen, iron oxidized to ferric state.
Genetics: Autosomal dominant.

B. The Inherited Structural Variants

1. **Single amino acid substitutions** in the α, β, γ, and δ-globin
chains
Molecular defect: Point mutations causing either
    a. normal amino acid change, e.g., HbC $\beta^{6\ Glu \to Lys}$; HbE $\beta^{26\ Glu \to Lys}$
    b. chain termination mutants
    c. nonsense mutations
    d. frameshift mutations

2. **Amino acid deletions**
Molecular defect: Deletion of 1 to 5 codons.

3. **Amino acid insertions**

4. **Sickle cell anemia**
Molecular defect: Abnormal hemoglobin: Hb S
    a. single amino acid substitution in β chain: 6 Glu ⟶ 6 Val
    b. electrophoretic mobility change
    c. Hb S is less soluble in deoxygenated state
Pathway affected: Blood flow and oxygen affinity.

C. <u>Quantitative Disorders of Globin Synthesis: The Thalassemias</u>

1. Abnormalities in the amounts of the different chains synthesized

2. Globin chains synthesized are normal in structure.

3. Molecular defect in globin synthesis include:
   a. gene deletion
   b. mRNA processing defects due to intron mutation
   c. nonsense mutations due to single base change
   d. nonsense mutations due to frameshift
   e. termination codon mutations.

4. α-Thalassemias: deficiency of α chain synthesis.
   a. homozygous for α thalassemia, all 4 α globin genes: death *in utero*, Hb Barts
   b. heterozygous for α thalassemia, 3 globin genes affected: Hb H disease
   c. $α_1$-thalassemia, 2 α globin genes affected
   d. $α_2$-thalassemia, 1 α globin gene affected.

5. ß-Thalassemias: decreased production of ß-chain, $ß^+$; absent synthesis of ß-chain, $ß^°$. Increased levels of Hb F and Hb H2.
   a. homozygous for ß-thalassemia: thalassemia major, Cooley's anemia. Severe microcytic anemia.
   b. heterozygous for ß-thalassemia: thalassemia minor.

X. COLLAGEN METABOLISM

1. **Ehlers-Danlos syndrome, types I, II and III**
   Molecular defect: Increased collagen fibril diameter.
   Genetics: autosomal dominant.

2. **Ehlers-Danlos syndrome, type IV**
   Molecular defect: Diminished synthesis of type III collagen. Another form has structural mutation in pro $α_1$(III)
   Genetics: Autosomal recessive.

3. **Ehlers-Danlos syndrome, type V**
   Molecular defect: Deficient lysyl oxidase.
   Genetics: X-linked recessive.

4. **Ehlers-Danlos syndrome, type VI**
   Molecular defect: Deficient lysyl hydroxylase.
   Diagnosis: Decreased hydroxylysine content.
   Genetics: Autosomal recessive.

5. **Ehlers-Danlos syndrome, type VII**
   Molecular defect: Structural mutation in pro $α_2$(I) chain.
   Diagnosis: Decreased procollagen aminoprotease activity.
   Genetics: Autosomal recessive.

6. **Osteogenesis imperfecta, type I**
   Molecular defect: Reduced type I/type III collagen.
   Genetics: Autosomal dominant.

7. **Osteogenesis imperfecta, type II**
   Molecular defect: Decreased synthesis of type I collagen.
   Genetics: Autosomal recessive.

8. **Osteogenesis imperfecta, type III**
   Molecular defect: Decreased secretion of type I procollagen associated with excess mannose in carboxyterminal propeptide.
   Genetics: Autosomal recessive.

9. **Cutis laxa**
   Molecular defect: Deficient lysyl oxidase with abnormal copper metabolism.
   Genetics: X-linked recessive.

XI. TRANSPORT DISORDERS

1. **Hartnup disease**
   Molecular defect: Intestinal and renal transport of amino acids, especially tryptophane.
   Diagnosis: Large quantities of amino acids, indolylacetic acid and other indoles excreted.
   Genetics: Autosomal recessive.

2. **Cystinuria**
   Molecular defect: Defective transport of cystine, lysine, arginine and ornithine.
   Pathway affected: The renal tubule and the GI tract.
   Diagnosis: Cystine crystals in urine; cystine stones in urinary tract.
   Genetics: Autosomal recessive.

XII. BLOOD COAGULATION: DEFICIENCY OF CLOTTING FACTORS

Disorder	Factor	Genetics
Afibrinogenemia	Fibrinogen	Autosomal recessive
Dysfibrinogenemia	Fibrinogen	Autosomal recessive
Parahemophilia	V	Autosomal recessive
Factor VII deficiency	VII	Autosomal recessive
Classic hemophilia	VIII	X-linked recessive
Christmas disease	IX	X-linked recessive
Stuart-Prower	X	X-linked recessive
Factor XI deficiency	XI	Autosomal recessive
Hageman Trait	XII	Autosomal recessive
von Willebrand	von Willebrand	Autosomal recessive
Fletcher trait	Prekallikrein	Autosomal recessive

XIII. ERYTHROCYTE ENZYMES

Enzyme Defect	Pathway	Genetics
Hexokinase	Glc $\longrightarrow$ G-6-P	Autosomal recessive
Pyruvate kinase	PEP $\longrightarrow$ pyruvate	Autosomal recessive
PK I isozyme affected	+ ATP	
Phosphohexose isomerase	G-6-P $\longrightarrow$ F-6-P	Autosomal recessive
Phosphotriose isomerase	DHA-P $\longrightarrow$ G-3-P	Autosomal recessive
Phosphoglycerate kinase	1,3-DPG $\longrightarrow$ 3-PG	X-linked
G-6-P dehydrogenase	G-6-P $\longrightarrow$ 6-PG	X-linked
6-PG dehydrogenase	6-PG $\longrightarrow$ Rbl-5-P	Autosomal recessive

XIV. TREATMENT OF INHERITED METABOLIC DISORDERS

1. Supply the missing protein, e.g., Factor VIII in hemophilia.

2. Deplete the stored substance, e.g., chelate copper with BAL or penicillamine in Wilson's disease.

3. Supply the missing metabolite, e.g., thyroxine in familial goiter (cretinism).

4. Limit intake of the precursor, e.g., galactose-free diet in galactosemia.

5. Avoid specific drugs, e.g., primaquine (antimalarial) in glucose-6-phosphate dehydrogenase deficiency.

6. Organ transplant, e.g., liver transplant in Wilson's disease; bone marrow transplant in ADA deficiency.

XV. LIST OF ABBREVIATIONS

1,3-DPG	1,3-Diphosphoglyceraldehyde
2,8-DHA	2,8-Dihydroxyadenine
3-PG	3-Phosphoglyceraldehyde
3ß-HSD	3ß-Hydroxysteroid dehydrogenase
6-PG	6-Phosphogluconate
8-HA	8-Hydroxyadenine
AdoCbl	Adenosylcobalamin
ApoA-1	Apolipoprotein A-1
Arg	Arginine
ASA etc.	Aryl sulfatase A, B, etc.
Asp	Aspartic acid
Cer	Ceramide
CP	Carbamoyl phosphate
DHA-P	Dihydroxyacetone phosphate
DHEA	Dehydroepiandrosterone
DHT	Dihydrotestosterone
F-1,6-diP	Fructose-1,6-diphosphate
F-1,6-diPase	Fructose-1,6-diphosphatase
F-1-P	Fructose-1-phosphate
F-6-P	Fructose-6-phosphate
Fuc	Fucose
G-1-P	Glucose-1-phosphate
G-3-P	Glyceraldehyde-3-phosphate
G-6-P	Glucose-6-phosphate
Gal	Galactose
Gal-1-P	Galactose-1-phosphate
Glc	Glucose
GlcNAc	N-Acetylglucosamine
Gly	Glycine
GMP	Guanosine monophosphate
Hb	Hemoglobin
HDL	High density lipoprotein
Hex A	Hexosaminidase A
His	Histidine
Ile	Isoleucine
IMP	Inosine monophosphate
LDL	Low density lipoprotein
Leu	Leucine
LP	Lipoprotein
Man	Mannose
MeCbl	Methylcobalamin
NAD	Nicotinamide adenine dinucleotide
OHCbl	Hydroxycobalamin
PEP	Phosphoenolpyruvate
Phe	Phenylalanine
Pi	Inorganic phosphate
PKU	Phenylketonuria
PRPP	Phosphoribosyl pyrophosphate
Rbl-5-P	Ribulose-5-phosphate
Ser	Serine
Tyr	Tyrosine
UDP	Uridine diphosphate
Val	Valine
VLDL	Very low density lipoprotein

XVI. REVIEW QUESTIONS ON HUMAN GENETICS

DIRECTIONS: Each of the questions or incomplete statements below is followed by four or five suggested answers or completions. Select the one that is BEST in each case and fill in the corresponding space on the answer sheet.

1. Overproduction of δ-aminolevulinic acid is characteristic of:

A. Multiple sulfatase deficiency.
B. Acid ceramidase deficiency.
C. Porphyria.
D. None of the above.

2. Lesch-Nyhan syndrome is due to a deficiency of:

A. Hypoxanthine-guanine phosphoribosyl transferase.
B. Phenylalanine hydroxylase.
C. Galactose-1-phosphate uridyl transferase.
D. Orotidine 5'-phosphate decarboxylase.
E. Neuraminidase.

3. Deficient fructose-1-phosphate aldolase isozyme B causes:

A. Hartnup disease
B. Galactosemia
C. Tarui's disease.
D. Hereditary fructose intolerance
E. Congenital lactase deficiency.

4. A receptor-mediated process is NOT involved in:

A. I-cell disease.
B. Familial hypercholesterolemia
C. Complete testicular feminization.
D. Tangier disease.

5. Deficient activity of glucose-6-phosphatase activity will result in:

A. Inability to metabolize fructose.
B. Abnormally high hepatic storage of glycogen.
C. Accumulation of abnormal glycogen in tissues.
D. None of the above.

6. Maple syrup urine disease is the result of a deficiency of:

A. Methylmalonyl CoA mutase.
B. Glucose-6-phosphate dehydrogenase.
C. Phosphoglucomutase.
D. Branched chain α-ketoacid decarboxylase.
E. Histidase.

7. Sickle cell anemia is the result of:

A. Decreased synthesis of α-globin chains.
B. Decreased synthesis of β-globin chains.
C. Single amino acid substitution in β-chains.
D. Frameshift mutation in β-chains.

8. Which of the following genes is on the X chromosome?

A. Fibrinogen.
B. Factor V.
C. Factor VII.
D. Factor VIII.

9. Sphingomyelinase activity is deficient in:

A. Niemann-Pick's disease.
B. Metachromatic leukodystrophy.
C. Krabbe's disease.
D. Gaucher's disease.

10. Glucose-6-phosphate dehydrogenase deficiency is:

A. Inherited in an autosomal recessive mode.
B. Most commonly manifested as a drug-induced hemolytic anemia.
C. Best treated by administering penicillamine.
D. Common in glycogen storage diseases.

DIRECTIONS: Each group of questions below consists of five lettered headings followed by a list of numbered words or statements. For each question, select the one heading that is most closely associated with it and fill in the corresponding space on the answer sheet. Each heading may be selected once, more than once, or not at all.

Questions 11-15:

A. Accumulate excessive lactic acid in the circulation.
B. Excrete excessive dermatan sulfate and heparan sulfate in the urine.
C. Be most benefited by bone marrow transplantation.
D. Excrete excessive methylmalonic acid in the urine.
E. None of the above.

11. A patient with Vitamin B_{12} deficiency will:

12. A patient with fructose-1,6-diphosphatase deficiency will:

13. A patient with Schie's disease will:

14. A patient with adenosine deaminase deficiency will:

15. A patient with Sly's syndrome will:

Questions 16-20:

A. UDP-glucose : galactose-1-phosphate uridyl transferase
B. Phenylalanine hydroxylase
C. β-Glucocerebrosidase
D. Glucose-6-phosphatase
E. Homogentisic acid oxidase

16. von Gierke's disease is caused by a defect in:

17. The most common form of galactosemia is due to a deficiency of:

18. Conversion of phenylalanine to tyrosine is catalyzed by:

19. Gaucher's disease is caused by a defect in:

20. Alkaptonuria is caused by a defect in:

Questions 21-24:

A. Cyanosis
B. Ehlers-Danlos syndrome, type V
C. Quantitative disorder of globin synthesis
D. Albinism
E. None of the above

21. Methemoglobin

22. Deficient tyrosinase activity

23. Deficient lysyl oxidase activity

24. Thalassemia

XVII. ANSWERS TO QUESTIONS ON
HUMAN GENETICS

1. C

2. A

3. D

4. D

5. B

6. D

7. C

8. D

9. A

10. B

11. D

12. A

13. B

14. C

15. B

16. D

17. A

18. B

19. C

20. E

21. A

22. D

23. B

24. C

12. NUTRITION

Thomas Briggs

I. MAJOR NUTRIENTS

A. Energy Nutrition

The unit generally used is the **Kilocalorie** (popularly but incorrectly also known as the Large Calorie): amount of energy needed to raise the temperature of 1 Kg water from 14.5° to $15.5^\circ C$. The megajoule is gaining some acceptance: 1 MJ = 239 Kcal; 1 Kcal = 4.2 KJ.

The usual figures for the energy yield from the metabolism of fuels are:

Carbohydrate	4 Kcal/g	Fat	9 Kcal/g
Protein	4 Kcal/g	Ethanol	7 Kcal/g

Respiratory Quotient (RQ) is the volume of CO_2 produced / volume of O2 consumed. This varies depending on the type of fuel being oxidized: 1.0 for carbohydrate, to 0.7 for fat.

Basal Metabolic Rate (BMR) is the rate of oxygen consumption, or equivalent heat production, of an awake individual, at rest, who has not eaten for at least 12 hours. To compensate for size differences, the BMR is usually expressed per unit surface area. A typical figure might be 35 $Kcal/hr/m^2$. It is higher in children, males, hyperthyroidism, etc.

Specific Dynamic Action (SDA) is an energy wastage or overhead, related to inefficiencies in metabolism, particularly of protein, and may be about 6% on a normal mixed diet.

The **daily requirement** of energy varies greatly, depending on BMR and especially on muscular activity. For a sedentary 70Kg man a typical value may be 2400 Kcal, but strenuous activity could double this.

B. Fuels

1. **Carbohydrates**: the major dietary form is the polysaccharide, **starch**. Important oligosaccharides include **sucrose, lactose**. Monosaccharides occur but are quantitatively less significant. Carbohydrates function almost solely as fuels.

2. **Fats** (lipids): **triacylglycerols** (TG) are the major dietary form. Other important lipids are **cholesterol, choline** (as phosphatidyl choline or lecithin, a source of methyl groups that can partially spare the methionine requirement). Dietary essentials include the unsaturated fatty acid, **linoleic acid** (18:2;9,12). Linolenic acid (18:3;9,12,15) may also be an essential but this has not been rigorously proven in humans.

Although fats have a major function as **fuels** and in the **storage of energy**, they also serve in the structure of **membranes** (phosphoglycerides, cholesterol), in the formation of **prostaglandins** (unsaturated fatty acids), as thermal **insulators**, and even for **decoration** (deposits of fatty tissue).

The amount and type of fat in the diet influence the levels of cholesterol in the body. A diet with a **high ratio of polyunsaturated to saturated fatty acids**, but low in total fat, tends to promote **low levels of cholesterol** in the blood.

C. Protein

Though proteins consumed in excess of the daily requirement are simply used as fuels, the most significant dietary·function is as a **source of amino nitrogen** for synthesis of body constituents.

Ten amino acids are considered dietarily essential because the human cannot synthesize the carbon skeleton:

Phenylalanine	Threonine	Histidine*
Valine	Isoleucine	Arginine*
Tryptophane	Methionine	Leucine
		Lysine

*A dietary requirement has not been rigorously established for arginine and histidine in the <u>adult</u> human.

The **quality** of dietary proteins depends on (1) digestibility and (2) the ratio of essential amino acids. The biological value (BV) depends mainly on amino acid composition. A more useful index, **Net Protein Utilization** (NPU), considers both (1) and (2) above. The NPU of human milk is 95%; of cow's milk, 81%; of wheat protein, 49%; of corn protein, 36%. Cooking can increase the NPU, as can mixing proteins with complementary amino acid compositions. This is especially important in vegetarian diets.

The **nitrogen balance** is zero when intake just equals loss (as urea, digestive losses, etc.). For this to occur, all essential amino acids must be present in the diet in sufficient quantities. If even one is deficient, negative balance results. Growth and convalescence are accompanied by positive nitrogen balance; illness, fever, starvation by negative balance.

D. Other

1. **Fiber**: non-digested (especially plant) material such as cellulose, lignin, etc. Though not generally regarded as a dietary essential, fiber nevertheless has beneficial effects on digestion. Through its **bulking** action, it promotes good mechanical functioning of the digestive tract; by a **speeding** action, it reduces the transit time for intestinal contents and therefore the time available for bacteria to produce possible carcinogens; by **binding** bile salts it increases the turnover of the bile salt pool and thus promotes excretion of cholesterol through increased conversion of cholesterol to bile acids. Fiber may, however, **decrease absorption** of certain nutrients, such as iron and calcium.

2. **Water**: requirement is highly variable. It functions as a **solvent** for components of blood and tissues, and as a medium for excre-

tion of wastes.   Another important use is in **regulation of body temperature.**

E. Health-Related Issues

1. **Recommended Dietary Allowance**: a figure with political overtones since many public health measures are tied to it. It is the amount of a nutrient which, if consumed by every member of a population, will keep nearly everyone in good health.   Since people vary greatly in their dietary requirements, no one figure is applicable to all. The RDA is based on an average requirement (which may not be known with precision), two standard deviations are added, and often a safety factor besides, so that it is a **generous excess** for most.   Detailed tables may be subdivided according to gender, age, etc. It should not be confused with a minimum requirement.

2. **Obesity** is a pervasive health problem related to energy intake because it predisposes to numerous conditions including cardiovascular disease, diabetes, and gallbladder disease.

3. On a world-wide basis, **malnutrition** is also a severe problem. There are two "pure" manifestations of malnutrition: **marasmus**, where total caloric intake is deficient, and **kwashiorkor**, where calories are sufficient but intake of protein is deficient. A real situation may be a combination of these.

4. For diseases related to deficiency of specific nutrients, see the appropriate sections to follow.

II. MICRONUTRIENTS

A. Vitamins

1. Fat-Soluble: are isoprenoid derivatives with varying degrees of unsaturation.

a. **Vitamin A**: is derived from a pro-vitamin, ß-carotene, a yellow plant pigment which is cleaved to vitamin A by an enzyme in intestinal mucosa.

i. Structure: occurs in three forms: **retinol**, vitamin A aldehyde (**retinal**), vitamin A acid (**retinoic acid**).

ii. Function: not known with precision except in vision, in which retinal is combined with the protein, opsin, to form the visual pigment, rhodopsin. The process of vision depends on the reversible conversion of 11-cis-retinal to all-trans-retinal. Further functions of vitamin A (for which a dietary supply of any form will suffice) are in differentiation of epithelial cells and in growth.

iii. Occurrence: ß-carotene in green and yellow vegetables, carrots, pumpkin, some melons. Vitamin A itself occurs only in animal products, especially butter (but often there by fortification), eggs, fish liver oil, liver (polar bear liver may contain toxic amounts).

iv. Deficiency: night blindness is an early warning. This may advance to keratinization of epithelial tissues, especially of the eye, causing xerophthalmia, a form of blindness which is of major concern in some areas.

v. Requirement:* 0.8-1.0 mg (as retinol), or six times that amount as ß-carotene, since the conversion is inefficient.

b. **Vitamin D**

i. Structure: derived from **7-dehydrocholesterol** through cleavage by UV light of the 9,10-bond, then rotation of ring A 180° around the 6,7-bond.

ii. Function: Cholecalciferol is like a prohormone. The active form is generated by (1) insertion of a hydroxyl group at C-25 by the liver, and (2) 1α-hydroxylation by a kidney enzyme whose activity is enhanced by parathyroid hormone. The active **1α,25-dihydroxycholecalciferol (calcitriol)** acts in a manner similar to that of the steroid hormones, to cause synthesis of a calcium-binding protein by intestinal cells. The effect is to raise the level of blood calcium and to promote mineralization of bone.

iii. Occurrence in foods is not necessary if a person has a minimal exposure to sunlight. Otherwise, dairy products, especially fortified milk, and fish liver oils are good sources.

iv. Deficiency: **rickets**. Adult form is **osteomalacia**.

* In this chapter, requirements are stated as the RDA for adults.

v. Requirement: **5-10 μg** (in the absence of sunlight).

c. **Vitamin E: tocopherols,** α, β, etc., depending on the length of the isoprenoid side-chain.

i. Structure:

ii. Functions as an **antioxidant,** probably by virtue of its ability to act as a trap for free radicals.

iii. Occurs widely, especially in vegetable oils.

iv. Deficiency: in animals, sterility and muscular dystrophy. In humans, anemia and neurological disorders may occur rarely.

v. Requirement: **8-10 mg/day,** but may be greater with increased consumption of polyunsaturated oils. These, however, are also good sources of vitamin E.

d. **Vitamin K (phylloquinone)**

i. Structure: a family of substituted naphthoquinones. The unsubstituted menadione is also active.

ii. Function: needed for the carboxylation of glutamyl residues to produce γ-carboxyglutamic acid in active blood **clotting factors** VII, IX, IX, and prothrombin. These γ-carboxyglutamic acids are able to chelate calcium. They also occur in other tissues, but with unknown function.

iii. Occurrence: ubiquitous in foods, and also synthesized by **intestinal bacteria.**

iv. Deficiency is rare but can occur in the newborn infant and in malabsorption syndromes, causing **hemorrhage** due to hypoprothrombinemia. Dicoumarol and related compounds are antagonists of Vitamin K and are used to prevent thrombosis.

2. Water-soluble, Energy Releasing

a. Thiamin (Vitamin B₁)

i. Structure: a substituted pyrimidine joined to a substituted thiazole.

ii. Cofactor Form and Function: **Thiamin Pyrophosphate (TPP)**. Part of the **pyruvate dehydrogenase** complex, TPP functions in the **decarboxylation** of pyruvate to form acetyl CoA. It is also part of α-**ketoglutarate dehydrogenase** and of **transketolase**.

iii. Occurrence: meat, liver, whole grains, vegetables.

iv. Deficiency: the classical disease is **beriberi**. Alcoholics may have a multiple B-vitamin deficiency known as **Wernicke's disease**.

v. Requirement: **1-2 mg/day**.

b. Riboflavin (Vitamin B₂)

i. Structure: a heterotricyclic system joined to ribitol.

FLAVIN

ii. Cofactor Form and Function: flavin **mononucleotide** or **FMN** (riboflavin phosphate); **flavin adenine dinucleotide** or **FAD** (a combined nucleotide with AMP). These, as part of flavoproteins, act in **transfer of hydrogen and electrons** from NAD to CoQ, from succinate, from acyl CoA in ß-oxidation, etc. FAD is part of **pyruvate** and α-**ketoglutarate dehydrogenases**.

iii. occurrence: milk, liver, meat, green vegetables. Cooking or exposure to light tends to destroy it.

iv. Deficiency: lesions of the lips, skin, genitalia.

v. Requirement: 1-2 mg/day.

c. **Nicotinamide, Nicotinic Acid (Niacin)**

i. Structure: a substituted pyridine.

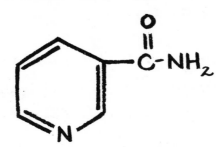

ii. Cofactor Form and Function: **Nicotinamide Adenine Dinucleotide** or **NAD** (nicotinamide + adenine + 2 ribose + 2 phosphate), **NADP** (NAD + a third phosphate). These act as **carriers of hydrogen and electrons** in a multitude of dehydrogenase reactions. In general, NAD acts in catabolism while NADP acts in synthetic reactions. NADPH and O_2 are used in mixed function oxidases, particularly in metabolism of drugs and in various hydroxylations.

iii. Occurrence: meat, liver, peanuts, legumes, whole grains. Nicotinate can be biosynthesized from tryptophane, but inefficiently (1 mg from 60 mg). Corn is a poor source of both niacin and tryptophane.

iv. The deficiency disease is **pellagra**, characterized by diarrhea, dermatitis, dementia, and death ("4D's"). Pellagra is of historical interest as it used to occur in a broad geographic area where the population was particularly dependent on corn as a dietary staple.

v. Requirement: 15-20 mg/day.

d. **Pyridoxine (Vitamin B₆)**

i. Structure: a substituted pyridine.

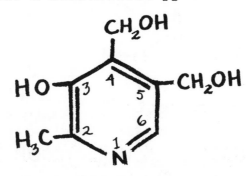

    ii.   Cofactor Form and Function: **pyridoxal phosphate** and **pyridoxamine phosphate**.   These act as coenzymes of **amino acid transaminases**, some **decarboxylations**, **ALA synthase**, etc. They reversibly form a Schiff base with amino groups.

    iii. Occurrence: milk, meat, liver, whole grains.

    iv. Deficiency: convulsions, oxalate kidney stones. Deficiency can be caused by treatment with the drug isoniazid.

    v. Requirement: 2 mg/day.

  e. **Pantothenic Acid**

    i. Structure: pantoic acid + ß-alanine.

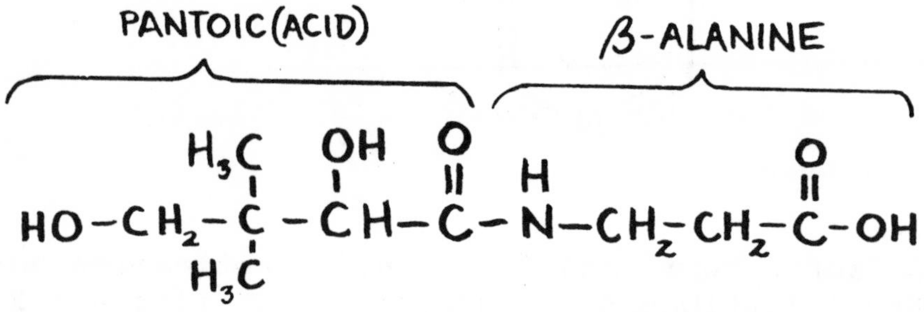

$$\underbrace{HO-CH_2-\underset{\underset{H_3C}{|}}{\overset{\overset{H_3C}{|}}{C}}-\underset{\overset{|}{OH}}{CH}-\overset{\overset{O}{||}}{C}}_{\text{PANTOIC(ACID)}}-\underbrace{N-CH_2-CH_2-\overset{\overset{O}{||}}{C}-OH}_{\text{ß-ALANINE}}$$

    ii. Cofactor Form and Function: **Coenzyme A** (phosphoadenosine diphosphate attached to the pantoic acid moiety, + thioethanolamine attached to the ß-alanine moiety). The terminal -SH combines with acyl groups to form a "high-energy" thioester; this is the way **acyl compounds are activated** before undergoing metabolism; example: acetyl CoA. The pantothenic acid and thioethanolamine are also part of **acyl carrier protein**, in which the -SH carries the growing chain during fatty acid synthesis.

    iii. Occurrence: ubiquitous in foods.

    iv. Deficiency: extremely rare.

    v. Requirement: 4-7 mg/day.

  f. **Biotin**

    i. Structure:

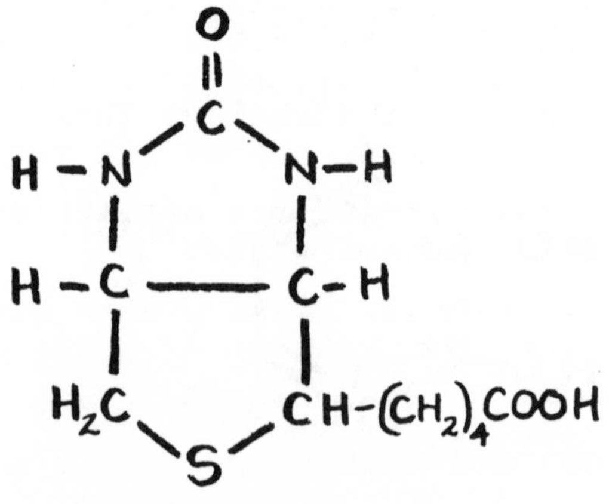

    ii. Cofactor Form and Function: biotin acts in several **carboxylations**, being a part of **pyruvate carboxylase** and **acetyl CoA carboxylase**, among others. It is linked to its enzymes by an ε-amino group of lysine, forming a swinging arm which enables it to transfer a -COOH from one site to another in the enzyme complex.

    iii. Occurrence: in many foods, and is made by intestinal bacteria in amounts sufficient to satisfy needs.

    iv. Deficiency: very rare; can be induced by consumption of large amounts of raw egg whites which contain a protein, **avidin**, which binds biotin and renders it unabsorbable.

    v. Requirement: **150-300 µg/day**, supplied entirely by **intestinal flora.**

3. **Water-Soluble, Hematopoietic**

   a. **Folic Acid** (pteroyl glutamic acid, folacin)

     i. structure

    ii. Cofactor Form and Function: **tetrahydrofolic acid (THFA)**. Active in metabolism of **one-carbon** units, THFA carries a C-1 fragment, at various states of oxidation, between N^9 and N^{10}. THFA is essential in **purine** synthesis, in the reversible transformation of **serine to glycine**, and in other aspects of one-carbon metabolism. The transported form of the coenzyme is N^5-methyl THFA; regeneration to active THFA requires adenosyl cobalamin, one of the forms of Vitamin B_{12}.

    Analogues of folic acid are useful as anticancer agents: **aminopterin, amethopterin, methotrexate.**

    iii. Occurrence: **kidney, liver, dark green leafy vegetables** (hence the name).

    iv. Deficiency: **macrocytic anemia.**

    v. Requirement: **0.4 mg/day.**

b. **Cobalamin (Vitamin B₁₂)**

i. Structure: a **corrin** derivative, a complex tetrapyrrole similar to the porphyrins, containing **cobalt**.

ii. Cofactor Form and Function: as the **deoxyadenosine** or **methyl derivative.** (Cyanocobalamin is a form often obtained from isolation procedures). **B₁₂** functions in two areas: (1) regeneration of **active tetrahydrofolic acid** (during which a methyl group is transferred to homocysteine to form methionine); (2) **mutase** reactions, as in the conversion of methylmalonyl Co A to succinyl Co A.

iii. Occurrence: only in foods of animal origin, especially meat and dairy products.

iv. Deficiency: all the symptoms of folate deficiency plus neurological abnormalities. In **pernicious anemia**, the deficiency is really one of **intrinsic factor**, a glycoprotein produced by the stomach, which is necessary for absorption of **B₁₂**. **Methylmalonic aciduria** is one effect.

v. Requirement: **3 µg/day.** But unlike other B-vitamins, B12 can be stored; the liver can hold a several years' supply.

4. Other Water-Soluble: **Ascorbic Acid (Vitamin C)**

i. Structure: related to the six-carbon sugars.

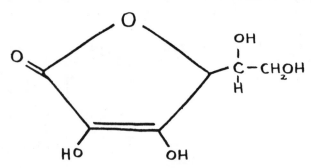

ASCORBIC ACID

ii. Cofactor Form and Function: cofactor (if any) not known. Vitamin C functions as a **reducing agent,** and is important in reducing dietary iron to the more easily absorbable ferrous form. It also is essential in some **hydroxylations,** especially of **proline** and **lysine** in **collagen** synthesis. It also probably has a function (unknown) in the metabolism of the adrenal cortex.

iii. Occurrence: citrus fruits, tomatoes, other fruits.

iv. Deficiency: The classical disease is **scurvy,** which is particularly a disease of collagen formation: poor wound healing, fragile bones, loosening of teeth, hemorrhage. Though blood levels of ascorbic acid decline rapidly on a deficient diet, scurvy usually takes many weeks to develop. Some authorities maintain that a state of deficiency short of outright scurvy is a very common condition.

v. Requirement: **60 mg/day.**

B. Minerals

1. Electrolytes

a. **Sodium**: The major **extracellular cation**, sodium functions in **osmotic, water, and acid-base balance**. It is ubiquitous in foods, especially those of animal origin and in preserved, prepared, and even frozen foods. A deficiency is uncommon, but can occur in an unacclimatized person through prolonged heavy sweating. An excess is widespread and insidious, and may cause **hypertension** in susceptible individuals.

b. **Potassium**: The major **intracellular cation**, potassium also functions in osmotic, water, and acid-base balance. It occurs widely in foods, especially those of plant origin. Deficiency may occur after prolonged vomiting, in diarrhea, diabetics, and in those on diuretics. Cardiac abnormalities, possibly fatal, may ensue. Excretion of sodium and potassium by the kidney is regulated by the **renin-angiotensin system**.

c. **Chloride**: The principal **anion** for Na^+ and K^+, it occurs everywhere and deficiency is not a problem.

2. Other Major Minerals

a. **Calcium**: the major inorganic element, calcium is 1.5-2% of the body's mass. Most is in the **skeleton**, but calcium has vital physiological roles in **muscle contraction, blood clotting**, as an intracellular **messenger**, and in many other ways. Dairy products are especially good sources. A primary dietary lack is rare, but deficiency can occur secondary to vitamin D deficiency, in women after multiple pregnancies and lactation (osteomalacia), and occasionally in a newborn on cows' milk (which has plenty of calcium but a low Ca/P ratio, which makes regulation difficult).

The level of blood calcium is regulated by **parathyroid hormone** (raises level), by **vitamin D** (promotes absorption in intestine), and **calcitonin** (lowers level). Bone serves as a buffer or reservoir. The adult requirement is **0.8-1.2 g/day** but adaptation may make the true need much less. Post-menopausal women and the aged have an increased requirement.

b. **Phosphorus**: an important **intracellular anion** (as phosphate), phosphorus is about 1% of the body's mass, occurring mostly as a constituent of **bone**. Chemically it has major functions as a component of **nucleic acids**, many **nucleotides**, and **phosphate esters** of sugars and other intermediates. A dietary deficiency is unknown, but low levels may occur in the diabetic. Requirement: **0.8-1.2 g/day**.

c. **Magnesium**: after potassium, an important intracellular cation. It occurs in **bone**, and also functions in many enzyme activities, especially those involving ATP. Most foods of vegetable origin are good sources, especially whole grains. Deficiency may occur in alcoholics. Requirement: **300-400 mg/day**.

3. Iron:

Functioning in oxygen transport and storage, iron is part of the heme proteins **hemoglobin** and **myoglobin**. It is also part of the **cytochromes** of the electron transport chain. Other heme-containing enzymes are **catalase** and **peroxidase** which are involved in the metabolism of H_2O_2. The average adult has about 4 g, 2/3 of which is in hemoglobin, and 1/4 in storage, particularly as liver ferritin. The phases of iron metabolism are outlined below:

a. Absorption: of dietary iron, only about 10% (1-2 mg/day) is absorbed, as the **reduced**, or **ferrous** form.

b. Regulation: is at the level of **absorption**. In response to adequate body iron stores, **intestinal cells** contain a **ferritin trap**, which prevents further passage of absorbed iron into the bloodstream before the cells slough off. When there is a need for iron, ferritin is not produced by intestinal cells, and iron that is taken in is allowed to pass through into the circulation.

c. Transport: iron is tightly bound by **transferrin**, two ferric atoms per molecule.

d. Storage: mostly in **ferritin** of liver, some in bone marrow. Many ferric irons per molecule.

e. Oxidation: ferrous iron is oxidized to the ferric form by **ferroxidase (ceruloplasmin)**, a **copper**-containing enzyme.

f. Excretion: virtually non-existent. The only way for the body to be significantly depleted of iron is by **bleeding** or **childbearing**. Iron overload may occur in some alcoholics, recipients of blood transfusions for hemolytic anemia, and overzealous practitioners of self-medication. Hemochromatosis is an inborn error in which too much dietary iron is absorbed, leading to excessive accumulation.

g. Occurrence: **meat** (heme iron is well-absorbed), egg yolk, legumes. Milk and spinach are poor sources. Simultaneous intake of **vitamin C** enhances the efficiency of absorption by helping to keep the iron in the reduced state.

h. Deficiency: iron-deficiency anemia is relatively common since dietary iron is inefficiently absorbed. It is most often seen in children, young mothers, and in persons with chronic loss of blood.

i. Requirement: **10 mg/day** (18 for teenagers and for women of childbearing age).

4. Trace minerals (most are required in the µg/day range)

a. **Iodide**: required for **thyroid** function-- T4 and T3. Found in sea foods, iodized salt. Requirement: **150 µg/day**. Deficiency is manifested as endemic **goiter**.

b. **Zinc**: important as a cofactor for many enzymes, e.g., **carbonic anhydrase**, alcohol dehydrogenase, nucleic acid polymerases. Deficiency is rare, but has been seen in alcoholics and in cases of renal disease. Requirement: **15 mg/day**.

c. **Fluoride**: a requirement is not established, but 1 ppm in the water supply is beneficial in helping to prevent **dental caries**.

d. **Copper**: part of a number of enzymes, notably **cytochrome oxidase** and other oxidases including ceruloplasmin. Wilson's disease is an inborn error involving excessive deposition of copper in the liver.

e. **Manganese:** widely distributed in vegetable foods, manganese participates in a variety of enzymic activities.

f. **Molybdenum:** required for **xanthine oxidase**, aldehyde oxidase, sulfite oxidase.

g. **Selenium:** functions in **glutathione peroxidase**, and as an **antioxidant** it is complementary to vitamin E.

h. **Chromium:** involved with the action of **insulin**.

i. **Cobalt:** the only need is as a component of **vitamin B_{12}**.

III. SUMMARY OF NUTRIENT - FUNCTION ASSOCIATIONS

Vitamin	Cofactor Form	Association or Function
A	11-cis Retinal Other forms	Rhodopsin in vision Growth; epithelial tissues
D	1α,25-Dihydroxy- Cholecalciferol	Intestinal calcium-binding protein
E	none	Antioxidant
K	none	Carboxylation of clotting factors
Ascorbic acid	none	Reducing agent; hydroxylations
Biotin	none	Carboxylations
Cobalamin	Deoxyadenosyl or methyl deriv.	Mutase reactions; recovery of active THFA
Folic acid	Tetrahydro deriv.	Metabolism of 1-Carbon units
Niacin	NAD NADP	Dehydrogenases, catabolic Dehydrogenases, anabolic
Pantothenic acid	Coenzyme A Acyl Carrier protein	Activation of acyl groups Growing chain in fatty acid synthesis
Pyridoxine	Pyridoxal-P	Transaminases, decarboxylases
Riboflavin	FMN, FAD	Dehydrogenases, catabolic
Thiamin	Thiamin PP	Pyruvate, α-KG decarboxylases

Mineral	Association or Function
Calcium	Bone; coagulation; muscle contrac; second messenger
Chlorine	As chloride, anion for Na^+, K^+, etc.
Chromium	Potentiation of insulin
Cobalt	Component of vitamin B_{12}
Copper	Cytochrome oxidase, ferroxidase
Fluorine	Prevent dental caries
Iodine	Component of thyroid hormones
Iron	Hemoglobin, myoglobin, in cytochromes of e^- transport, catalase, peroxidases
Magnesium	In bone; kinase reactions
Manganese	Arginase, acetyl CoA carboxylase
Molybdenum	Xanthine oxidase, etc.
Phosphorus	As phosphate, in bone, DNA, RNA, nucleotides, organic phosphates
Potassium	Intracellular cation; osmotic, acid-base balance
Selenium	Glutathione peroxidase
Sodium	Extracellular cation; osmotic, acid-base balance
Zinc	Carbonic anhydrase, alcohol dehydrogenase, etc.

IV. REVIEW QUESTIONS ON NUTRITION

DIRECTIONS: Each of the questions or incomplete statements below is followed by five suggested answers or completions. Select the one that is BEST in each case and fill in the corresponding space on the answer sheet.

1. Which of the following can completely replace dietary methionine?

A. betaine
B. cysteine
C. homocysteine
D. vitamin B₆
E. ornithine

2. Which of the following is most deficient in corn meal?

A. valine
B. methionine
C. threonine
D. tryptophane
E. leucine

3. A zero nitrogen balance is most likely in:

A. a normally growing child
B. a normal adult
C. a child on a diet low in tyrosine
D. an adult convalescing after surgery
E. a fasting adult

4. Which of the following is NOT a dietarily essential amino acid?

A. methionine
B. phenylalanine
C. leucine
D. isoleucine
E. serine

5. Raw egg white contains a protein, avidin, which:

A. digests fibrin clots
B. acts like trypsin inhibitor
C. produces a protein deficiency when fed
D. prevents absorption of biotin
E. blocks the formation of choline from betaine

6. The highest amount of ATP per gram is yielded by:

A. sucrose
B. starch
C. glutamic acid
D. oleic acid
E. succinic acid

7. Which of the following is impaired in biotin deficiency?

A. synthesis of ketone bodies from pyruvate
B. formation of lactate from glucose
C. synthesis of fatty acids
D. oxidation of fatty acids
E. elongation of fatty acids

8. Adenosyl cobalamin is involved in the conversion of:

A. malonyl CoA to acetyl CoA
B. α-ketoglutarate to succinyl CoA
C. Acetyl CoA to HMG CoA
D. propionyl CoA to methylmalonyl CoA
E. methylmalonyl CoA to succinyl CoA

9. Methotrexate is an effective antileukemic agent because it inhibits:

A. glycolysis
B. dihydrofolate reductase
C. thymidylate synthetase
D. glucose-6 phosphate dehydrogenase
E. RNA polymerase

10. 1,25-Dihydroxycholecalciferol (calcitriol) is the active form of

A. vitamin A
B. vitamin B
C. vitamin C
D. vitamin D
E. vitamin E

11. The three D's of dementia, diarrhea, dermatitis, are associated with a deficiency of:

A. vitamin A
B. vitamin B_{12}
C. vitamin C
D. biotin
E. niacin

12. Which of the following has a structure related to that of vitamin K?

A. Coenzyme Q
B. histidine
C. thyroxine
D. Christmas factor
E. vitamin E

13. Which of the following cofactors does NOT require a specific dietary vitamin precursor?

A. FAD
B. NAD
C. ATP
D. Coenzyme A
E. ACP

14. Parathyroid hormone increases the blood level of:

A. Mg^{++}
B. Cu^{++}
C. Ca^{++}
D. Fe^{++}
E. Na^{+}

15. Periodic injections of vitamin B_{12} can alleviate and prevent recurrence of the symptoms of pernicious anemia. These must be repeated monthly for

A. 3 months
B. 1 year
C. 2 years
D. 5 years
E. a lifetime

16. Wernicke's disease may follow a deficiency of

A. thiamin
B. riboflavin
C. niacin
D. folic acid
E. cyanocobalamin

17. In general, animal proteins are nutritionally better than vegetable proteins because

A. though they have no important difference in amino acid composition, they are more easily digested than vegetable proteins
B. the content of essential amino acids more closely matches the body's needs
C. they contain sufficient carbohydrate residues
D. vegetable proteins lack asparagine
E. vegetable proteins contain factors which inhibit protein synthesis

DIRECTIONS: For each of the questions or incomplete statements below, ONE or MORE of the answers or completions is correct. On the answer sheet fill in space

A if only <u>1, 2, and 3</u> are correct
B if only <u>1 and 3</u> are correct
C if only <u>2 and 4</u> are correct
D if only <u>4</u> is correct
E if <u>all</u> are correct

FILL IN ONLY ONE SPACE ON YOUR ANSWER SHEET FOR EACH QUESTION

Directions Summarized				
(A) 1,2,3 only	(B) 1,3 only	(C) 2,4 only	(D) 4 only	(E) All are correct

18. Starvation results in

1. decrease in liver phospholipid
2. increase in urinary acetoacetic acid
3. early breakdown of muscle protein
4. decrease in liver glycogen

19. In the mammal, tryptophane is a precursor of

1. NAD
2. Nicotinate
3. NADP
4. FAD

20. Utilization of propionate involves which of the following vitamins?

1. B_{12}
2. Pantothenic acid
3. biotin
4. folic acid

21. A derivative of folic acid is coenzyme for the:

1. formation of serine from glycine
2. formation of purines
3. metabolism of 1-C units
4. formation of urea

22. Oxygenases which replace C-H by C-OH require:

1. NADPH
2. tetrahydrofolic acid
3. O_2
4. H_2O_2

23. Ferritin

1. is a plasma protein which binds iron
2. is a protein which stores ferric ions
3. is involved in absorption of vitamin B_{12} from intestine
4. helps regulate absorption of iron in intestine

24. Absorption of iron from the intestine is increased by

1. giving ferrous (Fe^{++}) rather than ferric (Fe^{+++}) iron
2. administration of bicarbonate
3. hemorrhage
4. hemosiderosis

25. Long-standing biliary obstruction may cause malabsorption which can cause vitamin deficiency leading to

1. osteomalacia
2. beriberi
3. night blindness
4. megaloblastic anemia

26. Human dietary requirements include

1. methionine
2. cytosine or uracil
3. linoleic acid
4. glutamic acid

DIRECTIONS: Each group of items below consists of lettered headings followed by a set of numbered words or phrases. For each numbered word or phrase, select the ONE heading that is most closely associated with it and fill in the corresponding space on the answer sheet. Each heading may be used once, more than once, or not at all.

Questions 27-32: A deficiency of:

    A. ascorbic acid
    B. nicotinic acid
    C. thiamin
    D. vitamin D
    E. vitamin K

results in:

27. Pellagra.

28. Beriberi.

29. Hemorrhagic disease of the newborn.

30. Scurvy.

31. Hypoprothrombinemia.

32. Rickets.

Questions 33-37: A cofactor derived from

    A. thiamin
    B. pantothenic acid
    C. nicotinic acid
    D. pyridoxine
    E. biotin

is involved in:

33. Carboxylation of propionate.

34. Transamination of amino acids.

35. Transketolase.

36. Activation of palmitic acid.

37. Oxidation of lactate.

Questions 38-41: A deficiency of

    A. vitamin A
    B. zinc
    C. iron
    D. folate
    E. iodide

will cause:

38. Megaloblastic anemia.

39. Hypochromic anemia.

40. endemic goiter.

41. xerophthalmia.

Questions 42-45: The following pathways:

    A. tricarboxylic acid cycle starting from acetyl CoA
    B. glycogen synthesis from glucose
    C. gluconeogenesis starting from pyruvate
    D. glycolysis leading to lactate
    E. phosphogluconate oxidative pathway

require cofactors derived from:

42. Thiamin and nicotinamide.

43. Nicotinamide only

44. Nicotinamide, thiamin, riboflavin, and pantothenic acid

45. Nicotinamide and biotin

Questions 46-49:

    A. transaldolase
    B. methylmalonyl CoA mutase
    C. α-ketoglutarate decarboxylase
    D. cytochrome oxidase
    E. acetyl CoA carboxylase

46. Biotin

47. Adenosylcobalamin

48. Iron and copper

49. Thiamin

Questions 50-53:

    A. pyridoxal phosphate
    B. biotin
    C. thiamin pyrophosphate
    D. vitamin B_{12}
    E. tetrahydrofolic acid

50. Transaminase

51. Pernicious anemia

52. Pyruvate carboxylase

53. transformylase

Questions 54-57:

    A. cobalt
    B. molybdenum and copper
    C. copper only
    D. manganese
    E. zinc

54. Carbonic anhydrase

55. Vitamin B_{12}

56. Ferroxidase

57. Xanthine oxidase

V. ANSWERS TO QUESTIONS ON NUTRITION

1. C	21. A	41. A
2. D	22. B	42. E
3. B	23. C	43. D
4. E	24. B	44. A
5. D	25. B	45. C
6. D	26. B	46. E
7. C	27. B	47. B
8. E	28. C	48. D
9. B	29. E	49. C
10. D	30. A	50. A
11. E	31. E	51. D
12. A	32. D	52. B
13. C	33. E	53. E
14. C	34. D	54. E
15. E	35. A	55. A
16. A	36. B	56. C
17. B	37. C	57. B
18. C	38. D	
19. A	39. C	
20. A	40. E	